KU-166-441

'A delightful and insightful trip through gut health through the eyes of a medical chef! A spicy educational treat to be savoured: a delight.' **Professor Tim Spector, author of *Spoon-Fed***

'This is a book in the finest tradition of narrative recipe writing. It's a heavenly mix of whimsy, life and science, grounded in solid technique and blissful flavour.' **William Sitwell, food writer and restaurant critic**

'Saliha is amazing, this book is amazing. What we have here is the passion of an incredible cook, coupled with the knowledge of someone who has studied intensively the effects of our food on our health. This is not a diet book, but it is packed full of very useful, often incredible, and even fun topics, as well as super yummy recipes, as you would expect from a *MasterChef* champion. This book does so much, it will let you cook recipes like a top chef and at the same time give you the scientific foresight of a doctor.' **Gregg Wallace, *MasterChef* judge**

'A revelation and education for your taste buds, gut bugs, and dinner table conversation.' **Lisa and Alana Macfarlane, founders of *The Gut Stuff***

'This is an extraordinary fusion of science, literature, medicine and cookery. We've never read anything quite like it – a book that will transform your understanding of what you eat and how it makes you feel. The recipes work, are delicious, and many are utterly new and fresh. A perfect offering.' **Dr Xand and Dr Chris van Tulleken**

'I'm a big fan of Saliha – her recipes are always smash hits and I love how she writes and her way of weaving family memories, food history and kitchen storytelling in with medical evidence and the latest research. This is a great book for anyone who wants to cook a very tasty supper that hits the spot and also get to know their bodies, moods and emotions better. *Foodology* is both fascinating and full of delicious meals to enjoy cooking.' **Melissa Hemsley, author of *Eat Green* and *Eat Happy***

About the Author

Dr Saliha Mahmood Ahmed is a Specialist Registrar in Gastroenterology. She is the winner of *MasterChef* UK 2017 and author of *Khazana*, which won the Observer Food Monthly Best Cookbook of the Year 2019. *Khazana* was also shortlisted for the Travel Cookery Book of the Year in the 2019 Edward Stanford Travel Writing Awards and was winner of the Summer Harvest Gourmand World Cookbook Awards 2020, in the category Celebrity Chef in English. Saliha was the winner of Best Chef at the British Restaurant Awards 2019. She is based in London, works for the NHS (National Health Service), and has two young children.

Photograph © Manjit Riyat

A FOOD LOVER'S GUIDE TO DIGESTIVE HEALTH & HAPPINESS

by Dr Saliha Mahmood Ahmed

with a foreword by Sat Bains

The information in this book is not intended to replace or conflict with the advice given to you by your GP or other health professional. All matters regarding your health should be discussed with your GP or other health professional. The author and publisher disclaim any liability directly or indirectly from the use of the material in this book by any person.

First published in Great Britain in 2021 by Yellow Kite
An Imprint of Hodder & Stoughton
An Hachette UK company

3

Page Design by Briony Hartley
Illustrations by Rachael Tremlett © Hodder & Stoughton 2021
Illustration on page 152 based on a Shutterstock image

Image on page 130 is from *The Science of Spice* by Dr Stuart Farrimond
© Dorling Kindersley Limited. Published by Dorling Kindersley Limited and reproduced with their permission.

A CIP catalogue record for this title is available from the British Library

Hardback ISBN 978 1 529 31982 8
eBook ISBN 978 1 529 32300 9

Typeset in Magneta and DIN by Goldust Design

Printed and bound in Great Britain by Clays Ltd, Elcograf S.p.A.

Hodder & Stoughton policy is to use papers that are natural, renewable and recyclable products and made from wood grown in sustainable forests. The logging and manufacturing processes are expected to conform to the environmental regulations of the country of origin.

Yellow Kite
Hodder & Stoughton Ltd
Carmelite House
50 Victoria Embankment
London EC4Y 0DZ

www.yellowkitebooks.co.uk

Contents

Foreword

Food is an obsession for me and has been for as long as I can remember. When I met Saliha at the *MasterChef* finals, she was recreating one of our chocolate desserts, and a background in the sciences was all too evident in her working methods and the way she approached the technical dish. She nailed it and went on to win.

As a professional chef for over thirty-five years, I can attest that we use many of the subjects covered here in the restaurant when creating new dishes, flavour combinations and textures. This book gives you insight and advice from a professional gastroenterologist whose obsession is food – the perfect combination, which you can see in the delicious recipes here. This incredibly insightful and comprehensive book does not just help people who have an interest in food. I will recommend it to all my chefs at the restaurant as part of their understanding and ongoing quest around the behaviour of all our senses and texture, and how they play a vital role in the pleasure of eating.

The sight, the smell, the texture, the odours of food are crucial to understand, as well as balance and digestion. As a chef, I will come back to this book over and over again for reference. A lot of work is covered here – from gut health and digestion, to texture, sound and

language and how important they all are to the overall subject of food, giving you a better understanding of how we eat.

The way Saliha has interwoven anecdotes and personal references with recipes and stories also makes this easy to read. You will definitely be hungry at several points throughout! If you are interested in food in any way at all, this book is a must. It will educate and enhance your understanding of what food is and how it is digested, elevated and made delicious with some fantastic recipes I can't wait to try.

Sat Bains,
Two Michelin Star Chef at
Restaurant Sat Bains

In loving memory of my paternal grandmother,
Naseem Akhtar, for showing me how an unrelenting
positive attitude can alter the trajectory of your life.

Introduction

DIGESTIVE HEALTH AND HAPPINESS – AND WHY YOU NEED IT

As a society, we have a growing tendency, almost an obsession, to analyse what we eat. Deciding what to consume for breakfast, lunch, dinner (not to mention any snacks in between) is, in some ways, far more stressful and complex than it has ever been. The messaging around what is 'best' to eat can be perplexing and anxiety-provoking for some, overwhelming for others and exhausting for many.

In this climate of gastronomic confusion there can also be social pressure pushing you to 'define' your own relationship with food. Are you the calorie-counting carbohydrate restrictor who fasts once a week? Or the vegan, wholefood-eating believer in nutritional supplementation? Perhaps you see yourself as the vegetarian – but occasional pescatarian – who also eats turkey for Christmas. Do you eat ready meals with pride, or with a heavy heart full of regret after a long and exhausting day at work? Maybe you always eat organic produce bought exclusively from the local farmers market; or perhaps you go to the local supermarket because you prefer identical-looking spherical tomatoes and enjoy the convenience. Do you fancy yourself a food critic, is fine dining your extravagant pastime or are you a connoisseur of all the curry houses within a five-mile radius of your sofa? It may be that food ethics and environmental concerns impact your food choices, or that the scale of these arguments, and the sheer number of possible solutions, leave you more confused than when you started ...

And this is by no means even *close* to an exhaustive list of potential food lifestyle choices. Try spending an hour on Instagram reading some of the captions written under the #instafood tag and you'll either end up chucking your phone against the wall in hashtag-fuelled desperation, or you'll find a culinary niche you never knew existed. But I have often found myself asking the question: is all this choice liberating or debilitating? Are we in the midst of a food revolution or a food catastrophe? And which, if any of these options bring us closer to finding a sense of digestive health and happiness?

Clearly, food is important. And its importance is no longer just limited to survival; food is now a pastime, an industry, a lifestyle, an object of celebration when we are happy and a thing of comfort when we are sad. It reaches into every part of our modern lives – from birth to death and everywhere in between.

From the food that we choose to put in our supermarket shopping trolleys to the money we spend, from the grocers we end up forming relationships with to the restaurants we choose to frequent, and perhaps most importantly, to our long-term physical health and mental wellbeing, the impact of food choices is utterly astounding. Which is why finding health and happiness through food is so fundamental.

As a gastroenterologist and cookery author, I meet many people who take a keen interest in trying to work out my food habits. Colleagues have tried to peer sneakily into my lunch box at work, curious about what exotic wonders it might contain. Maybe they expect a twelve-hour slow-cooked lamb, flavoured with spices that you can only order off the dark web, or that I will have replaced my plastic Tupperware with a traditional Moroccan earthen tagine. Those people are often left disappointed, because my lunch box might well contain just a slice of cold pizza and a slightly bruised-looking apple, even though I did win *MasterChef* UK in 2017.

On any given day, I might have sushi for a mid-morning snack, followed by a toastie for lunch and perhaps a curry for dinner. But I also find deep comfort and joy in preparing gloriously aromatic broths, fibre-rich salads and fermented foods galore. If you have four hours to spare, ask me about my kombucha (for those who don't know, this is a slightly vinegary tasting, fizzy, fermented green tea which tastes divine).

Every food has its place in life, and knowing that place is important. It's about being self-aware, intuitive, mindful even, around what you eat and when – knowing that you are eating because you have made an informed choice, free from external judgement or pressure. It isn't about following a prescriptive diet. Rather, it's a harmonious mindset towards food and eating, underpinned by the science of how we eat, taste, digest and process the food we choose to put through our bodies.

With eating having become so complicated in this day and age, I feel relieved that (most of the time) I have trained my brain and body to relate to food in a balanced way. I acknowledge that this can be quite tricky to achieve though, in a world where mixed culinary and scientific messages co-exist against a backdrop of vast food options.

Here is an example of the type of dilemma one may face on a day-to-day basis. Imagine you are really hungry, starving in fact. In front of you is a paper bag full of the very best jam doughnuts; soft, pillowy dough encased in crystals of sugar, bursting with tart, gushing raspberry jam. The stuff of dreams. What do you do? Not eat the doughnuts and save your appetite for a salad? Demolish the entire packet? Eat one doughnut and save the rest, reluctantly, for later? What is the conversation that goes through your mind in this scenario?

To simplify, you have found yourself in that classic 'naughty-but-nice'/guilty-pleasure food scenario. When I was in this position I ate – no, actually, devoured – a jam doughnut ... and it tasted every bit as glorious as you'd expect. Do I feel guilty about eating it? No, absolutely not. It was cooked to perfection and brought me great joy. We humans, by our very design, are attuned to enjoy sweet food; we possess taste buds that recognise sugar and feed into the reward pathways in the brain, making the lure of the doughnut hard to resist.

Do I also realise the harm eating doughnuts will do to my body? Yes, of course I do. Knowing the science behind how my blood sugar spikes when processed foods are absorbed rapidly by the gut and the biology of how hunger and food cravings work is sufficient deterrent to prevent me from bingeing on the remaining four doughnuts in the bag. Simply put, I understand the science behind why I relish the jam doughnut, and I also understand that it tells me to eat them in moderation. This may seem like common sense to some of you, but for many people it isn't that obvious; some may choose to abstain from the doughnut completely, whereas others find themselves polishing off the whole lot, all the while searching for the next sweet treat to feast on.

Reading through the pages of this book you will find two kinds of messaging. The first pertains to how the digestive system/human body is designed as a vehicle to enjoy food. These will help make you the very best possible cook you can be, hence laying down the groundwork for a sense of digestive health and happiness while teaching you to derive maximum fulfilment from food. In other words, an exploration – via how humans' taste, the role of umami, the world of spices, food textures and more – of why that jam doughnut tastes and makes you feel so darned good.

The second type of messaging will focus on what the most up-to-date science tells us is beneficial to eat. I will be outlining how it is possible to balance one's love for mouth-watering food with a desire to eat 'healthily', i.e. why munching through the entire pack of doughnuts does the digestive system no real favours. I delve into the phenomenon of hunger versus fullness, the role of our gut bugs and the concepts of bloating and constipation in delicious detail.

A gastroenterologist's journey to digestive health and happiness

One thing you'll notice as you progress through this book is that a lot of my personal journey towards digestive health and happiness is focused on the science of the gut. As a medical student, I was trained to understand the function of the bowel; cellular structures, gut physiology and intestinal anatomy were my language for many years. The teaching was not remotely glamorous, and even the most devoted science nerd such as myself was left underwhelmed and a little bored. But one thing that struck me was that throughout my learning, food and eating were never – and I really do mean never – mentioned.

By the time I had graduated, it was beginning to sink in how, by food not being brought up during my training as a gut doctor, my education on the human digestive system was fundamentally incomplete. It seemed more and more absurd every time I thought about it. For example, patients with bowel-related ailments would, understandably, ask me what they could eat to help their symptoms, but I was usually forced to direct the conversation away from food, shifting the focus to things like their medicines and symptoms – things that I had been trained to discuss. In some cases, I could refer them to a dietician colleague, but generally food discussions were not on the menu.

This was a wake-up call for me. Why, as doctors, were we not interested in what our patients ate? An individual's daily menu gives us a huge amount of valuable information about their health, and discussion of a patient's diet can produce tangible, long-term and, in some cases, even lifesaving improvements.

You see, the gut is the place where the food we eat undergoes the magical transformation to become part of us. It is one of those critical points of contact between the rest of the universe and the body, but it is often not given the credit it deserves, particularly in scientific circles.

It is staggering to think that every morsel of food ingested undergoes the most extraordinary of journeys. It meets muscles and bends, violently strong acid, a multitude of chemicals, hormones, nerves, bacteria and foreign organisms. What we need is extracted and what we can't use is expelled, re-entering the wider world (yes, as poo).

It was Hippocrates, the founding father of modern medicine, who first had the foresight to notice that all health begins at the gut, and that it can be impacted by the food we eat. What's more, he had this insight around 400 years before the birth of Christ. Yet somehow, it seems that his message has been lost somewhere over the succeeding millennia.

In the last ten years, Hippocrates has been quoted a lot in 'heathy eating' literature, but in fact, until relatively recently, the science underpinning gut health was in its infancy. Nowadays, with ever-growing numbers of studies revealing the most intriguing insights into the impact of gut health on our overall wellbeing, it is astounding to discover the extent of the relationship between our emotions, our physical environment and our choice of food. And once you've realised that the food you eat can have a tangible effect on your life, it's clear to see how your food choices can determine health and happiness.

As a doctor now working in the field of gastroenterology, I am able to see first-hand the impact of a negative relationship with food on a patient's broader health and sense of wellbeing. Whether it is obesity, with all the associated morbidity, or the (incredibly difficult to treat) obsession with weight loss and the quest for a perfect body, a person's relationship with food can affect many vital aspects of their life.

Through my work as a doctor, I have met people with nearly every food-induced symptom you can imagine – from bloating and bubbling and excessive flatulence, to troublesome belching, abdominal pain severe enough to make them bed-bound, nausea and vomiting, reflux, indigestion, heartburn, diarrhoea and the most inconceivably dreadful constipation.

Changes in diet, alongside conventional medical treatments, have helped alleviate some of these worrying gastroenterological issues, and to reignite a healthy relationship with food in these patients. And it is precisely this reason that I have chosen to write a book that gives equal credence to extraordinary-tasting food and the science of 'healthy' eating. In its very distilled form, you can see this book as a chef and doctor's manifesto on how to eat for both pleasure and health.

About this book

My research, both for my work and for this book, has led me to the sad conclusion that the various players in the food and digestive health sector often fail to speak to one another effectively. For example, the food-manufacturing industry doesn't talk much to the food researchers, while the clinical practice conducted by gastroenterologists remains at arm's length from the work of food scientists and gastrophysicists. Food journalists, cookery authors, professional chefs, food historians, nutritionists, dieticians and other allied disciplines should also have a part to play in weaving together the patchwork quilt of knowledge of the relationship between food and our bodies. I feel strongly that each discipline offers a fresh and valuable insight and I have referred to many of them in these pages. With so many players and perspectives, and with food being such an intimate part of our social fabric, I am resolute in my belief that food deserves to be an 'ology'. Food: a vast subject of study, a dynamic and evolving branch of knowledge, a discipline in its own right.

Ultimately, conversations with my patients, family and friends, the relationship with my own bowel and a love of eating are what prompted me to write *Foodology*. It is, without doubt, written for every reader to enjoy (not just those suffering with digestive disease or symptoms) and it may well be the first book to take you on a culinary journey that incorporates science, medical research and recipes, stopping at all the major points of interest from the mouth to the bum!

To be clear once again, this is not a book about 'dieting' in the conventional sense. Nor is it a manual on how to make each and every food decision you might be faced with. It is an unapologetic celebration of what I believe to be the most amazing organ of the body – the mind-bogglingly intricate digestive system –and what happens when certain foods are put through it. It is also a culinary anthology: a collection of carefully chosen anecdotes that relate to food and how it is perceived and processed by the body.

Interspersed throughout are the most glorious, mouth-watering recipes that inspire and awaken the taste buds. From those that showcase fermented foods and spices, to new ways of appreciating fibre-dense foods and much more, the recipes serve as examples of how you might incorporate the science I talk about into your day-to-day life, in your own kitchen, on your dining table, in your lunch box. By making these recipes part of my culinary repertoire, I have brought great joy to my loved ones ... and, most importantly, to myself. For those readers who feel that their affection for food is wavering, I am optimistic that these recipes will reignite their love for 'yummy' once again.

I hope that in time, each and every reader of this book will come to a sense of equilibrium with their eating, whatever route their journey takes. And the journey is important – because food and cooking are properly joyous. In its deeply sensorial, emotion-laden brilliance, food has the magical ability to comfort, cultivate and fulfil. Bringing more than simple nutrition and sustenance, food is vitality and wonder at the same time, it fashions growth and is the very fuel that powers life through our weird (and wonderful) bodies. Cooking and eating have been two of the most important ways in which humans have acquired a deeper understanding of the natural world that we inhabit, and without being too Mother Gaia about it all, food really is the strongest connection there is between the earth's biology and our own.

On that note, let us begin.

Roadmap of the gastrointestinal tract

Before we get into the main text, let's take a look at the basic parts and functions of the digestive tract.

1. **The oral cavity:** this is the point of entry for food, and it's also known as your mouth. Apart from tasting food, courtesy of a tongue lined by taste buds, the mouth is also where the mechanical and chemical digestion of food begins. The tongue and cheeks help create a ball of food, which is then swallowed. Once we have swallowed, a complex neuromuscular process called deglutition propels the formed food ball into the oesophagus.

   Chapters 2, 3, 4 and 5 look at some of the amazing things that go on in the oral cavity. Chapter 2 explores how we taste, and Chapter 3 focuses on a less well-known type of taste called umami. Chapter 4 looks at how texture affects our eating experience, in particular things that go 'crunch', while Chapter 5 looks at how our bodies respond to one of the most divisive of ingredients: chilli.

2. **The oesophagus:** this tube, approximately 20cm or so in length, connects the mouth to the stomach. Food doesn't just fall through the oesophagus, it moves downwards via a series of rhythmic contractions and propulsive movements called peristalsis. Watching peristaltic movements is like watching a Mexican wave pass through a crowd, and is why, if you were to do a headstand after taking a bite of a cheese sandwich, the food doesn't drop back into your mouth. The contractions push food in just one direction, towards the stomach. The oesophagus meets the stomach at a place called the gastro-oesophageal junction, which acts as a one-way barrier preventing acidic content from the stomach rising up into the oesophagus. We briefly visit the oesophagus in Chapter 5.

3. **The stomach:** this muscular, bean-shaped sac is the widest part of the digestive tract and has some serious storage capacity, shrinking when empty but expanding to almost twice its size when full. Cells lining the stomach secrete powerful acids, which aid digestion and are the first line of antimicrobial defence for the body. The stomach churns under subconscious control, macerating food and mixing it with a host of enzymes to form a mushy liquid called chyme. The stomach triggers the release of an orchestra of hormones, some of which help to control our appetite, while others manage the secretions of enzymes and gall-bladder function. Handily, it does not digest itself, as it is lined by a thick layer of mucus which protects its constantly regenerating lining.

   We look at the stomach in Chapter 6, which deals with how our bodies experience hunger and satiety, as well as the multitude of sensations and processes that make up what we refer to as our 'appetite'.

4. **The small bowel (also known as the small intestine):** this is the core processing centre of the digestive tract. It comprises around 6 metres of snake-like, coiled-up tubing and is where the majority of the body's digestion and absorption of nutrients take place. The inner part of the small intestine is lined with tiny hair-like projections known as villi; under a microscope these appear finger-like, making the surface of the small bowel resemble a shag-pile rug. The villi help to maximise the surface area available in the small bowel for nutrient and water absorption and, laid flat, it can almost cover a tennis court. It's ironic that we call it the 'small' bowel, given that it's so very humungous.

   The surface layer of cells in the small intestine is called the epithelium, and it secretes mucus that protects the lining from the acidic chyme entering it. The epithelium lining of the small bowel regenerates approximately every four days or so. Food moves through the small bowel by peristalsis, just like in the oesophagus.

5. **The large bowel (also known as the colon):** the undigested contents of the small bowel enter the large bowel through a one-way valve called the ileo-caecal valve. The first part of the colon is called the caecum, connected to which is an appendage called the appendix. Previously thought to be useless, the appendix is actually a reservoir for beneficial microbes, which help keep the colon teeming with microbes that live symbiotically within the gut (more on this later). The colon acts a bit like a fermentation tank, working on all the undigested bits the small bowel was unable to break down and generating gaseous by-products in the process.

   The colon is also responsible for absorption of water, along with the processing of a variety of vitamins (B and K) and bile salts. It rises up the right side of the body, across the top of the abdomen and then down the left, where it eventually transitions to the rectum and anus. Handily, the mushy stool entering the colon has become slightly more formed by the time it reaches the end of the colon and the rectum.

6. **The rectum and anus:** the rectum is a 10–15cm segment of bowel which functions as a storage chamber for your poo. In the West, we each make an average of around 200g of stool a day, consisting mainly of dead gut bacteria, fibre, intestinal cells and digested red blood cells. The stool is dispatched to the outside world through the anus, which is a 5–7cm-long ring of muscular tissue, in a process called defecation, otherwise known as 'pooing'.

   The latter part of the digestive system is made up of the large bowel, rectum and anus, and these parts are explored in more detail in Chapters 7, 8, 9 and 10. In Chapter 7, we look at our microbiome, the vast colony of bugs that we are now beginning to realise determines more about our mental and physical health than we ever imagined. Chapter 8 considers how this microbiome affects our brains through what's known as the gut–brain axis and what happens when things go wrong and the gut microbiome becomes imbalanced. Chapter 9 examines the phenomenon of bloating; and in Chapter 10 we look at look at the most common of gastroenterological ailments: constipation.

Organs of the digestive system

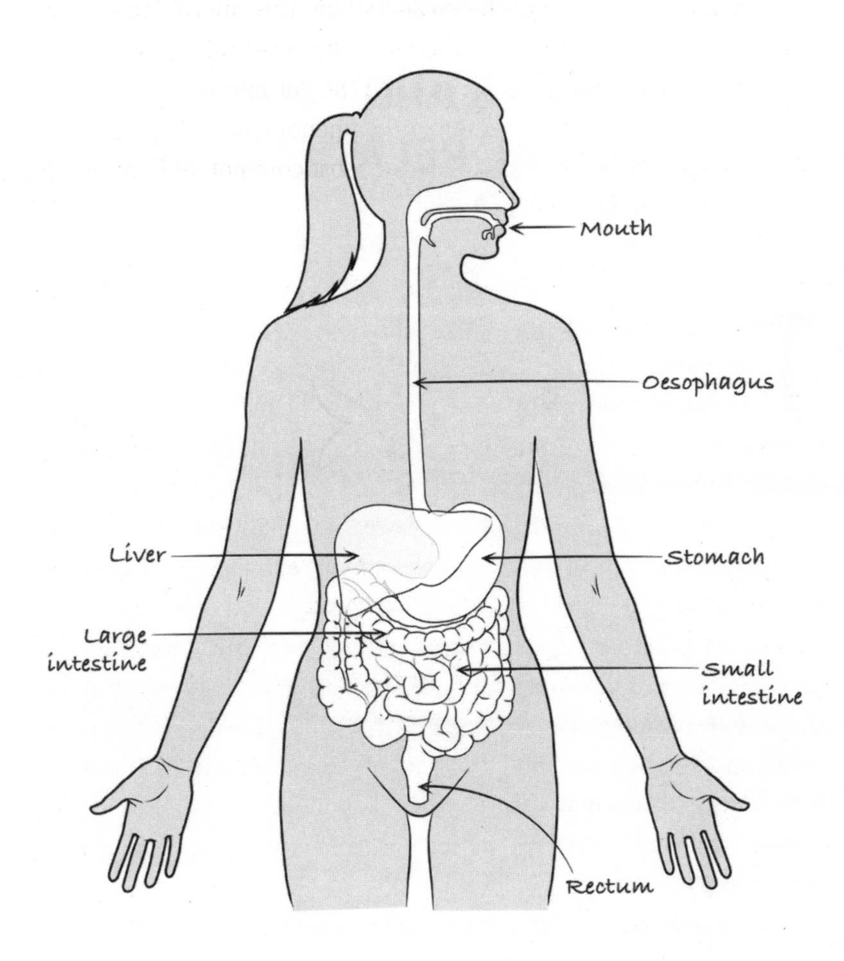

Chapter 1

HUMANS AND FOOD – A SPECIAL RELATIONSHIP

Throughout the years of my childhood, my mother was plagued by me with the question, 'Why?' Why did I need to eat my peas at dinnertime? Why were peas good for me? Where did peas come from? Why were peas green? Why did I not like the way peas tasted? The questions went on and on, and I simply would not eat my peas until I was given satisfactory answers to all my queries.

With the passage of time and development of a social conscience, I stopped asking the question 'Why?' so vocally. And instead I sought answers not just for *why* day-to-day things happened, but also *how* they happened. I would spend hours browsing through pages of my illustrated encyclopaedia, feeding my insatiable curiosity. I even recall being absorbed by its 'food and cookery' section, which explained how fruits and vegetables were grown, how flour was milled and how cheese was made. Perhaps an early sign of my future love of food and eating.

It was only when I was much older though, that I realised that the need to know 'why' and 'how' things work is an essential part of making sense of the world around us, whether in the context of a spiritual quest or scientific exploration. Thus, in a book that is a food lover's guide to bringing you closer to finding digestive health and happiness, I feel that a few fundamental questions must addressed from the outset. Firstly, why do we actually bother cooking? Secondly, how did we form our digestive system? And finally, when do we learn to taste and develop flavour preference? The answers to these questions are not just fascinating bits of knowledge, they actually form the very framework upon which a sense of digestive health and happiness is established.

'Coquo ergo sum' – I cook, therefore I am

If we're going to talk at all about finding digestive health and happiness, we need to start at the beginning by understanding the history of the human relationship with food, and the transformative role that cooking has played in unlocking the nutrients necessary for us to evolve into what we are today.

Cooking is a very human trait. Imagine if farmyard chickens cooked corn fritters for lunch, cows prepared a meadowsweet salad for brunch and garden foxes simmered scavenger's stews in their dens. Sounds strange, doesn't it? Yet it's not when we talk about humans doing it. In fact, all known human societies today, from the Arctic tundra to sub-Saharan Africa, cook their food. From an anthropological standpoint, cooking is a key activity that defines us as human beings and separates us from the wider animal kingdom.

Consider this menu:

- Breakfast: fibrous and bitter leaves, honey and fruit
- Lunch: bark, more fruit, raw monkey meat and brain tissue
- Dinner: leaves, grubs and a portion of antelope

Sound good? Probably not, but don't worry, it's not the latest fad detox diet. It is, in fact, a typical diet of one of our closest living relatives: the chimpanzee. However, it is extremely difficult to extract calories from this sort of raw-food-based diet, and as a result, chimpanzees must forage for food more or less continuously.

And it is not just chimps; gorillas and orangutans in the wild will spend around a spectacular six to eight hours a day simply eating. Raw food, like that served on the chimp's menu above, is harder to chew, digest and extract calories from than the cooked food we enjoy. As human beings, if we tried to extract enough calories from a chimp's diet to sustain our modern lives, we would not thrive.

Richard Wrangham, a leading British biological anthropologist who works at Harvard University, is credited with the invention of the 'cooking theory'. By virtue of being so deliciously provocative, it has become one of my favourite anthropological models. In its most simple form, it suggests that the ability to cook food has shaped human evolution more than almost any other human behaviour.

The story begins around 2 million or so years ago, when a major evolutionary change took place. *Homo habilis* began to evolve into *Homo erectus* (our most recent evolutionary forefathers) before the emergence of humans in their current form, *Homo sapiens*. *Homo erectus* fossils tell us that at the same time that the size of our ancestors' brains was increasing, the size of their teeth and digestive tracts was decreasing.

So what? What does a big brain and small teeth have to do with cooking?

Wrangham argues that when *Homo erectus* began cooking their raw ingredients (most likely on an open fire) they were able to extract more calories from the same quantity of food. And, as cooked food needs to be chewed less forcefully and is digested more easily, two things happened: fewer calories were spent consuming the same amount of food, and the teeth and digestive tract also became more specialised, decreasing in size.

As a cook, this makes sense to me. We know that beef carpaccio is cut paper thin because it would take ages to chew down on a big lump of raw beef, but barbecued tender beef ribs, where the meat is falling off the bone, practically slips down the throat without resulting in a sore jaw.

The higher number of extracted calories from cooked foods allowed the brain of the *Homo erectus* to grow larger, and it did so very quickly indeed. A human brain is a very hungry organ; it needs more energy for its size than any other organ in the body, and at any one time uses around a fifth of our total available energy. So, because our ancestors cooked food before they ate it and spent less of their energy digesting raw food, it ultimately allowed our brains to harness more energy from our diets, and to grow bigger and more complex.

Wrangham's theory is well supported, and although it is contested fervently by some, there is very persuasive evidence in favour of its claims. For example, in Wonderwerk, South Africa, anthropologists have unearthed signs of control of fire, perhaps for cooking, in caves 1 million years old. Around 400,000 years ago primitive hearths for cooking appeared, followed by earthen ovens around 250,000 to 300,000 years ago. Pots and cooking utensils from some 20,000 years ago have even appeared in archaeological sites in China. We now know that Neanderthal remains show burnt bones and spices and herbs adhering to their dental plaque, making me wonder what a Neanderthal barbecue tasted like.

Although the scientific community hasn't yet reached a unanimous agreement on exactly when the cooking of food started, it (whenever it began) clearly improves the quality of ingestible food and the amount of it that is available to us to use as energy (think cooked versus raw potato), or to use at all (as cooking will stop food perishing quite so quickly). What is also clear is that social behaviours around food must have changed profoundly after the advent of cooking. These changes could have manifested themselves in a range of helpful ways; for instance, instead of needing to travel from food patch to food patch, somewhere along the way our ancestors might have been able to accumulate food in a particular defined territory and wait with it until it was cooked to their desire.

Sadly, no time machine exists to either prove or disprove Wrangham's cooking theory. How I would have loved to have been a fly on the wall at the decisive point in human history when the very first alchemic reaction of cooking food took place. One small step for primitive man, one giant evolutionary step for mankind.

The term coctivor, from the Latin *coquere* (to cook) has been used to describe the human as one designed to cook what is eaten. As a chef and mother, I can't help but feel a sense of endearment towards the term. The process of cooking plants and meat has brought a multitude of joys to my life and those of my friends and family, and knowing that cooking is an activity which separates us from other earthly creatures has allowed me to also appreciate its significance for modern humans everywhere. In other words, *coquo ergo sum* – I cook, therefore I am.

Knowing that cooking is such an important part of the human experience has allowed me to lean into my love for cooking, and to embrace food in all its forms, from the most complex dishes that take days to prepare, to the humble beans on toast. After all, we owe a debt of gratitude to that first Neanderthal man or woman who picked up a smouldering branch after a lightning storm and thought, maybe I should put my steak on this. Without them, we might still be feasting on monkey brains.

By knowing where we came from and the role cooking has played in turning humans into the amazing form that we are today, I find myself yet more in love with this powerful, hugely influential craft. Cooking has shaped us and is a part of us all, and I hope you will agree that this is a key bit of knowledge in each of our quests to find our own individual sense of digestive health and happiness.

So, I would encourage those of you who are reluctant to step into the kitchen to give cooking a chance. I honestly believe that there is no one who 'just can't cook'. The beauty of cooking is that you can learn to do it at any stage of your life, and with the rising popularity of step-by-step videos and cooking tutorials online, the resources and guidance to help you become a competent home cook are vast. Whether you are leaving for university, wish to take up a new hobby or have lost the person in your life who made your food, becoming master of your kitchen and cooking, the most fundamental of human activities, will allow you to reap a multitude of rewards and is the first step in laying the foundation to forming a sense of digestive health and happiness.

How does a gut become a gut?

It's all well and good knowing about the history of the human race and its experiments into cooking, but (predictably as a doctor) what makes me feel even more at peace in my relationship with food is knowing how we, as individuals, develop into creatures that are designed to love and embrace it. And to learn more about this I had to delve into the complex medical field of embryology.

Embryology is the branch of biology that studies the development of the foetus. I used to feel overwhelmed when I had to attend embryology lessons at medical school; I always felt this sense of amazement, followed by complete confusion at how changes were taking place in the foetus. Even the language used to describe left and right, forwards and backwards, up and down was different: imagine a lecturer speaking at a hundred miles per hour in a language you don't understand, using words like medial, lateral, caudal, cranial, endoderm, ectoderm, gastrulation, topological structure, blastula ...

Granted, I chose to become a doctor, but still, it was all a bit much. I felt a bit like a child all over again, sitting a non-verbal reasoning exam, staring at rotating shapes, imagining how a blob of cells could eventually morph into a being that bears any resemblance to a fully formed human. I managed to cram enough to get by, but never really felt I had a grip on the subject.

Some years later, after leaving medical school, while working as a junior doctor, I came across the case of a young baby born with gastroschisis – a birth defect in which the baby's intestines extend outside of the abdomen through a hole next to the belly button. I had trained to control my emotions in front of patients and their families, but to face such a beautiful baby with this heartbreaking condition, alongside a wonderful, doting pair of new parents facing the biggest challenge of their lives pushed me to my limit. I was, at the time, six months pregnant, and a deep-rooted fear of how my baby was forming inside me sprang up like never before.

I realised that I needed to improve my understanding of the development of the digestive system. The need to intellectualise the experience was part of how I coped with it, so, I once again picked up those dusty embryology books, desperate to know where things had gone wrong for this innocent little baby.

The formation of the digestive system starts almost immediately, after a mere four weeks of gestation. This is a time when many mothers may not even know that they are pregnant. The baby, at this stage, is barely the size of a poppy seed, a little dot, measuring just 2mm in any direction. A disc of cells folds to form a blind-ended tube and divides rather neatly into three key regions. These are named, predictably, the foregut at the top, the midgut in the middle and the hindgut at the end, and all three have their own distinct blood supplies. Humans are deuterostomes, meaning that in embryonic development, the first opening to form is the anus, not the mouth. For those of you who are wondering what this means, it means that you started your life as an arsehole. No offence, just a fact.

The bit that I find particularly amazing is that this simple tube then manages to specialise its function, to form what I would argue is one of the most architecturally and functionally complex systems in our bodies. What's more, every organ in your body begins its life as a piece of your gut. The formation of individual organs with their specialised cell types happens when small 'outpouchings' develop from the gut tube. For example, in the foregut one outpouching develops into the liver, while another becomes the pancreas and another the gall bladder and bile ducts. Surprisingly, even the lungs form from the foregut tube. By thirteen weeks, organ formation, or 'organogenesis' of the intestine is complete.

When a baby is born, the average length of their intestine is an astonishing 275cm. Not only is the human intestine longer than any other mammal's, it is also far more mature at birth. For example, just halfway through pregnancy, the human gut cells responsible for the absorption of nutrients will be as mature (or more) as those of a suckling rat that has been alive for two weeks.

If you further examine the tiny miracle growing inside the uterus, you will find that the foetus makes swallowing motions from just eleven weeks onwards and sucking movements at eighteen to twenty weeks of gestation. With all these incredible developments taking place, I wondered, could my baby actually taste the food I was feeding it? Are babies forming their own relationships with food, finding their own digestive health and happiness, in our bellies?

Pre-natal flavour learning

I had cravings in my pregnancy like I have never had before. Some were transient, while others persisted throughout. For the full nine months of my first pregnancy, all I wanted to do was demolish platefuls of chips. My eyes would light up as I walked into our NHS canteen at lunchtime to see a hot batch being dropped into a metal tray, fresh from a deep-fat fryer – glistening under the golden lights of a commercial warmer, soft and fluffy-centred with a crisp exterior, covered with a generous sprinkling of salt. Oh, how I swooned.

You will be happy to know that I have returned to having a more measured love of chips since giving birth. And although I still adore *pommes frites* in all their wonderful forms, those untameable cravings have dissolved into mere nostalgia with the passage of time. But back then, it did make me wonder: do we learn to taste when in our mother's womb, or later on when we are weaned? Would my baby love chips because I loved them when I was pregnant?

Many people will impart their words of wisdom to a pregnant

woman (whether they have been invited to do so or not) on how they should eat to optimise their baby's growth. For example, a friend from Singapore who became anaemic once told me that she was encouraged by a great-aunt to eat red soil from a mound of termites to replenish her iron stores. In Japan, women are sometimes advised to avoid spicy food for fear of making the baby ill-tempered, while in Tanzania some women abstain from eating meat out of a traditional belief that the baby might take on the characteristics of that particular animal. I suppose, deep down, nobody wants to have a child with the personality of a goat.

And there are examples like this from around the globe, varying with culture and geography. The trouble is that most are steeped in ancient folklore, scary stories and lots of non-evidence-based superstition – closer to mystique and magic than to facts and medicine. So, instead of asking if my baby would love chips, maybe what I should have been asking was: can 'flavours' of food actually permeate the placenta and infuse the amniotic fluid which surrounds the baby and which he or she swallows?

When I think about the effect of food on the flavour of bodily secretions, my mind is drawn to a memorable scene from the once popular sitcom, *Sex and the City*. Samantha, our protagonist, is rather fond of a gentleman named Adam Ball, aside from one small detail: when having oral sex, she finds that his semen tastes like 'asparagus gone bad'. Samantha's friends have a giggle and decide, after much deliberation, that a good place to start would be to write to Martha Stewart about what Adam Ball could eat to make his spunk sweeter. Now I realise that this is a little crude for a book about cooking, but the point that this scene illustrates rather unambiguously is that the food we eat *can* eventually enter our bodily secretions. This is why after you eat beetroot your pee can turn purple, or after eating onions your sweat can start to smell a bit, well, oniony.

Handily, there are scientists who have studied this. One study compared the amniotic fluid of pregnant women who did and did not swallow garlic capsules. After forty-five minutes a sample of amniotic fluid was taken from each of the participants. People were then asked to sniff the fluid and see if they could detect the smell of garlic in it. Unsurprisingly, they were able to very reliably distinguish the garlic-infused from the non-garlic-infused amniotic fluid. Researchers now claim, with a high degree of confidence, that amniotic fluid is a rather complex 'first food' for the foetus and even has a smell and a taste associated with it.

In another experiment, the babies of mothers who drank carrot juice in their last trimester of pregnancy were filmed while being fed carrots for the first time, and the videos were compared with others of babies whose mothers who did not have any carrots at all in their diets. The results? The babies of mothers who had carrots in pregnancy were more likely to enjoy them when tasting them for the first time, with noticeably fewer negative facial expressions.

Similar experiments have been conducted with newborn babies of mothers who have and have not been exposed to anise during pregnancy. Yet again, the facial expressions of babies presented with an anise scent were much more likely to be positive among those whose mothers had been exposed to it.

The love of all things sweet seems to be innate, almost like we are programmed to swoon over them. Around fifteen weeks after conception, a foetus will actually demonstrate an appreciation for sugar by swallowing more sweet amniotic fluid than that infused with a bitter taste. The love for sweet things is so strong in childhood that when you give children sweet solutions of varying strengths and ask them to choose their favourite, they will tend to gravitate towards the most saccharine sweet one. In comparison, adults tend to choose a level of sweetness that they feel is right for them, with some revulsion towards the most concentrated sugar solutions.

For the very young, eating may be a very intense experience. It is thought that infants have around 30,000 taste buds in their mouth

...pregnancy cravings...

Food cravings are strong urges for foods that are more specific than mere hunger and very difficult to resist. The exact aetiology of cravings in pregnancy is not known. Some hypothesise that hormonal changes and increased smell and taste sensitivity may be responsible. However, overall data proving this theory remains scant and the exact nature of the link between smell, taste, hormones and pregnancy cravings has yet to be elucidated. Previously, it was thought that pregnancy cravings might be fulfilling some sort of nutritional deficiency. In actual fact, most are for foods which are not necessarily rich in any particular nutrient, the most popular one in the Western world being for chocolate.

Some research points to pregnancy cravings having a psychological source. Many of the foods craved – e.g. chocolate, fried chicken or macaroni cheese – are inherently pleasurable items of food, but are also associated with a sense of guilt, being 'forbidden' or 'naughty'. Some psychologists feel that cravings arise in pregnancy as a result of it being a time when it is socially sanctioned/acceptable to eat foods that women feel they are usually meant to stay away from.

There may be additional cultural factors at play driving cravings in pregnancy, which can be a demanding time, both physically and emotionally. Studies in Tanzania have noted that providing the desired/craved food can be a sign that the woman is getting the social support she needs from partner and their families. So, when those 2 a.m. cravings for a burger kick in, and your committed other half takes you to the nearest drive-through, it is not just the burger that is intensely pleasurable, it is the fact that someone you love has bought the burger for you; this has a value that extends way beyond the calories.

(the number is closer to 9,000 in adulthood). So, it is no wonder that children have a strong aversion to bitter foods and why nursery food is traditionally quite bland. Some scientists have even suggested that babies are born with 'synaesthesia', where senses are entwined with one another – so, children may taste food in 3D technicolour with associated food sounds being amplified...or so the theory goes.

Individually, taste buds go through constant cycles of regeneration lasting approximately a fortnight. This is why we recover the ability to taste after burning our tongues on hot beverages. The trouble is that as we age, the taste buds continue to die and shed, but fewer regenerate. With fewer taste buds in the mouth, food tastes more bland overall and requires more seasoning. In addition, a declining sense of smell, changes in the composition of saliva and multiple medications can combine to influence taste perception in later life.

For me, all this research makes me think about the choices I made in my pregnancy, and whether they will have any long-term impact on my son's eating habits. I shall admit now that he adores chips more than any of his peers. I still remember the glee on his face when he tried them for the first time, at six months old, mid-weaning with a dribble-laden, smiling face, gurgling with excitement, jerking all his limbs happily, while sitting on my kitchen counter. His fondness was immediate – something I hadn't seen with the other foods I had introduced him to.

However, I also ate vast quantities of cherries when I was pregnant with my son, and he seems to avoid them like the plague. So where does that leave us? Ultimately, the composition of amniotic fluid and maternal eating is very hard to research, there are only a few studies to refer to and even fewer extending past infancy into childhood and adolescence.

I do sometimes wonder if it would have been preferable for me to suppress my craving for chips in pregnancy, and whether this would have been better for my son in the long term. Maybe I should have exercised more restraint. But honestly, I find it to be a fruitless and potentially destructive task to look back in regret on choices that I felt right at the time (and if any of you thinks that you could have come between a pregnant woman and her craving, I applaud your naivete).

Our brains link flavours with experiences from infancy onwards. When a baby drinks its mother's milk it is simultaneously being held and comforted, and this is where the association with eating and comfort is thought to be born. As we grow older, our taste preferences can and often do change, but the effects of early food learning may linger. According to some scientists, the reason why vanilla ice cream remains the most popular flavour globally is that it is the closest in taste to breast milk. Why else would something so plain be the world's best-selling flavour?

The passage of time does have an impact on flavour perception. The unbridled love for sweet food usually makes way for a more sophisticated palate that enjoys vegetables, olives, strong cheese and all those other 'grown-up' tastes. An environment where children are encouraged to try a variety of different foods is likely to widen their repertoire of tastes significantly. When I think back to my childhood, I realise that many of the foods I love now were bitter torture as a child. Broccoli, grapefruit and dark chocolate come to mind, and I am sure this experience is one that I share with many of you. I guess the point I am making is this: just because your mother didn't eat sprouts when she was pregnant doesn't mean you won't love them as an adult. Flavours grow on you with time and practice.

For me, the fact that our individual relationship with food probably begins all the way back in the foetus is testament to just how fundamental the relationship between humans and food is.

Whenever I find a new favourite ingredient, or see my son find love for a new food, it cements the emotional connection I have with food. As a food lover, finding happiness, for me, is not just about understanding the science behind food. It's about understanding *why* that science matters, and it's about being able to identify the emotional and physical ways that it manifests itself in our day-to-day lives.

Summary

- My first piece of advice for all those wishing to achieve a sense of gastronomic happiness is to cook, cook and then cook some more. If you think you aren't great at it, that's ok! Take heart in the fact that even the Neanderthals could piece together a barbecue! We are biologically designed to eat cooked food, so don't be scared to step into your kitchen.

- I am not advocating we all eat cake every day, but it is important to realise that we are biologically inclined to enjoy sweet food. It is completely natural to desire it, particularly as a child, but moderation is key.

- The first food you tasted and swallowed was your mother's amniotic fluid. It is possible that the food a woman enjoys when pregnant and breastfeeding can influence early taste preferences.

- If your mother binged on crisps, sweets and chocolate while pregnant, this doesn't mean you will necessarily like any or all of these things – but you might. Taste preferences change over time.

Chapter 2

THE MOUTH AND TASTE BUDS: ENTER THE CAKE HOLE ...

Honestly, we aren't going to get very far in our quest for gastronomic happiness if we don't have a tête-à-tête about taste. Knowing that cooking made us what we are today, and that we are biologically designed around our ability to extract the most nutrition from food is great, but really, it's the taste and texture of what we eat (and maybe also how Instagrammable it is) that define our modern relationship with food.

With that in mind, I wonder if you can think of anything more satisfying than a truly special gastronomic experience? I'm not talking about that cold sandwich at lunch, or the quick breakfast of cornflakes and milk you shovel into your mouth before running out of the door. I'm talking about those all-consuming taste experiences that you can recall with perfect clarity just sitting here reading this paragraph – the ones that makes you salivate with a single thought. And as your tummy gurgles with longing, you are overcome with the urge to run into the kitchen and consume anything you can lay your hands on, if only to satisfy the craving for a few minutes.

We've all been there. I remember one of these experiences, quite vividly. When I was a child, maybe only five or six years old, my cousin Zain and I were dragged along with the rest of our family to visit the home of a distant relative. Zain and I were great friends, almost inseparable. We would spend hours playing doctors and patients, jungle wars, snakes and ladders and hide and seek. He was the ringleader, and I dutifully followed his orders (most of the time).

I remember it was autumn, maybe winter in the early 1990s, definitely raining outside. We must have been somewhere in the north of England, perhaps Yorkshire. I was proudly sporting a new red corduroy dress which was kept aside for special occasions. In Pakistani families, special occasions are the ones where different branches of the family come together to present their children to the group, demonstrating how well behaved they are and therefore competing for the prize of 'best child in the family' (which, let's be honest, really means 'best parent').

Zain and I sat at the dining table on our best behaviour, waiting for lunch to be served, both quite hungry after the long and uncomfortable journey in my parents' rattling little Lada. As children, we were, of course, very aware of where exactly the food was and how much of it was within our reach; so when the aroma of deep-fried pastry wafted from the kitchen, filling the air with the scent of celebration, and then, from above our heads, a plate landed in front of us, we knew without having to look that the samosas had arrived.

These were not any old shop-bought, defrosted samosas. They were the real deal. Heavenly golden triangles, homemade by an elderly relative who had made them a thousand times before, for a thousand family celebrations. I still remember every crunch of perfect savoury, crisp crust, each encasing the perfect amount of filling. Every mouthful gave you a substantial amount of spiced, fatty, moist lamb mince, tempered gently with the sweetness of bright green British garden peas. Next to the samosas, no less important for the overall experience, was a huge glass bottle of ruby-red tomato ketchup, into which we dunked our samosas joyfully and without restraint.

A few times, I caught sight of my mother giving me a death stare. It was the type of stare where her eyes said everything: *stop eating those samosas. Right. Now.* She did not want our relatives to think her daughter greedy, or that she was denying me the pleasures of good food or regular feeding. Greedy, ill-mannered children were often the subject of idle family gossip, and she went to great pains to ensure that her greedy, ill-mannered children were not the topic of conversation.

But these samosas were delicious, too delicious to heed even my mother's stare. We crunched away, ignoring all the other food on the table, all the while avoiding my mother's gaze, until, of course, the huge stack of samosas was reduced to one. Zain looked at me, and I looked at him. It was a moment of tension, the type that only comes about when two young family members compete over scarce but tasty resources. We were two cowboys, staring each other down on a sandy street in the Wild West, ready to draw. *A Fistful of Dollars* might as well have been playing in the background. Who was going to get the last samosa on the plate?

Being the youngest, I had prepared myself to dutifully give it up, but Zain looked at me benevolently. 'Go on, you eat it', he whispered. I remember grinning while devouring the last samosa, nibbling it away, corner by corner, thrilled that my cousin and childhood best friend would sacrifice it for me.

I am certain that no matter where you grew up, whether your family was big, small or somewhere in between, that you have a food memory as vivid as the one I've just described. Memories like these, that we have all constructed at some point in our lives, make up a patchwork of taste experiences imbued with a rainbow of emotions, and etched for ever in our minds. For me, these memories mature over time, gaining sentimental value each time I revisit them. To this day, I still maintain a very close and loving relationship with lamb samosas, and yes, they must still be dunked in ketchup.

But how much do we really understand about the science that underpins these memorable taste experiences? To put it bluntly, do we actually know how taste works?

Lamb and Sweet Pea Samosas

Serves 4

Ingredients

For the filling:
150g fatty lamb mince
1 heaped teaspoon cumin seeds
½ teaspoon garam masala
1 teaspoon ginger paste
1 teaspoon garlic paste
½ teaspoon paprika
2 teaspoons tomato purée
100g defrosted peas
3–5 green chillies, finely sliced
Salt, to taste

For the pastry:
150g plain flour
1 teaspoon nigella seeds (optional)
1 teaspoon salt
2 tablespoons light olive oil
Vegetable oil, for frying

Method

1. First prepare the samosa filling. Place the fatty lamb mince in a non-stick pan heated on a high heat. (If it is too low, then the mince will stew and release water rather than browning and releasing its fat.) Brown the mince in the frying pan and, using a wooden spoon, try to break the chunks of mince into smaller pieces that can later be easily stuffed into the samosa. It will take around 5–7minutes to brown your mince fully.

2. Now add the cumin seeds, garam masala, ginger and garlic pastes and paprika and turn the heat down to medium-low. This will prevent the spices from burning and becoming bitter. Finally add the tomato purée and 250ml warm water. Allow the mince to simmer for 15–20 minutes, or until the moisture has evaporated almost completely and the fat is starting to separate out. Season with salt to taste, then remove the pan from the heat and leave to cool.

3. Add the frozen peas and green chillies to your preferred level of spice. (I like samosas really hot, so I often opt for 5 green

chillies, but I appreciate this may be a little too much for some.) When the filling has cooled completely, it is ready to use.

4. To prepare the pastry, put the flour, nigella seeds (if using), salt and olive oil into a large bowl. Adding just enough water to form a dough, use your hands to bring the dough together. Knead gently until the dough no longer sticks to the bowl and then divide it into four small balls. Cover with clingfilm to prevent drying out. Leave to rest in the fridge for around 30 minutes, if you have time, or use immediately if you are in a rush.

5. Roll each dough ball out on a lightly floured surface into a circle that is less than a millimetre thick and has a diameter of about 20cm. Cut the circles in half to make two semi-circles. Brush a little water along the straight edge of one semi-circle and pick it up. Form into a cone shape by folding the two corners in so that they meet in the middle and one wet edge overlaps the other. Press the dough edges together to seal.

6. Fill the dough cone with the filling to about three-quarters of the way up. Brush the remaining flap of pastry with a little more water and seal by pinching the edges together with your fingers, or crimp using a fork. Continue with all the pastry and filling (you should have eight samosas).

7. Place the prepared samosas on a greased tray and cover with a damp cloth or clingfilm, so they don't dry out. You can make these up to a day ahead and chill in the fridge or freeze for another day; simply defrost for about 1 hour before frying.

8. When you are ready to serve the samosas, heat the vegetable oil in a large pan or deep fryer; it's hot enough when a small piece of pastry dropped in sizzles immediately. Deep-fry the samosas in small batches for about 90 seconds on each side (turn them carefully with a slotted spoon), or until they are golden brown

and crisp. Remove from the oil and drain on kitchen paper.
Serve hot with any pickle or relish of your choice (ketchup is my preferred condiment).

What makes taste, taste?

Whenever I think about which foods make me most happy, it's always the tastes that I remember comforting me when I was younger. But our relationship to taste, like so many things related to food, is equal parts science and emotion: equal parts *Why does that samosa taste good?* and *Why does the taste of that samosa make me feel good*? However, even though taste is one of the main interfaces to help us make sense of the world around us, the neuroscience and physiological understanding of it has remained in its infancy until quite recently.

From as far back as the 4th century BC, humans have known that there are a selection of basic tastes that we are able to discern: sweet, salty, sour, bitter. More recently, we have added the umami taste (more accurately the detection of the compound glutamate) to this list (see p. 68). Initially, these tastes were thought to be detected in specific areas of the tongue, with sweet receptors at the tip, bitter ones at the very rear and salty and sour receptors on the sides. It may not surprise you to learn that this simplistic version of reality, popularised by repetition and the passage of time, could not be further from the truth.

The mouth is the place where our most important taste buds live. Beautiful, complex constructs, these taste buds are column-shaped cells embedded in the surface of the tongue. Each human taste bud is not a single lone cell, but an organised cluster, made up of hundreds of taste cells shaped like garlic bulbs. At the base of the taste cells is a network of nerves, called sensory afferent nerve fibres, which take information away from the taste bud to higher centres in the brain, where the electrical nerve impulses can be translated into what we perceive as taste.

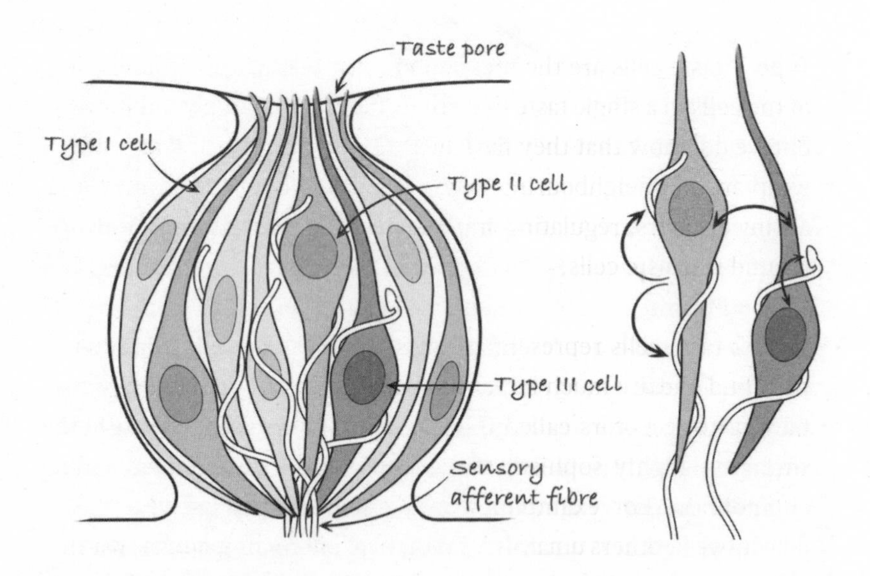

Taste trivia

As an interesting little fact to pull out at the next pub quiz: scientists have identified a number of taste buds in places that are very far away from our mouths – in our brains, hearts, stomachs, intestines, pancreas, lungs and, believe it or not, the testes. The function they serve in these locations remains largely a mystery, but I think we can all be happy that the taste buds located in the pancreas, stomach and testes are switched off. I cannot imagine it would be a particularly pleasant experience to taste whatever they have to offer, while snacking on a chocolate croissant.

In essence, you can think of taste buds as an elegant conducting device for 'gustatory stimulus' (also known as taste), and broadly speaking, scientists classify them into three types of taste cells, named Type 1, Type 2 and, wait for it ... Type 3.

- **Type 1** taste cells are the most numerous, making up around half of the cells in a single taste bud. Their function is largely unknown, but we do know that they have little finger-like projections which wrap around neighbouring taste cells. Type 1 cells may function as tiny cleaners, regulating and clearing the external environment around the taste cells.

- **Type 2** taste cells represent around a third of the cells in a single taste bud and are much larger and broader than Type 1 cells. They have taste receptors called G-protein-coupled receptors on their surface – highly sophisticated devices that detect certain taste compounds. For example, some G-protein-coupled receptors detect sweet, others umami and yet others bitter compounds. When a taste compound binds to the surface of a G-protein on a Type 2 taste cell, it causes the activation of a special protein inside the cell. This protein initiates the production of a barrage of molecules that activate the delicate nerve endings at the base of the taste buds.

- **Type 3** taste cells are, in contrast to Type 1 and 2, few and far between. They vary by region in the mouth, being much more prevalent in taste buds at the back of the tongue compared to the front. They detect sour tastes and are responsible for your love of lemon cordial, lime juice in your guacamole, pickled onions and Tangfastic Haribo. These slender, elegant cells do not work in the same way as Type 2 cells. The consensus is that Type 3 cells transduce acidic stimuli through at least two separate molecular mechanisms, thus informing our brains that there is something tart lurking around in our mouths.

You may have noticed that I have not touched on how salty tastes are decoded by taste buds. Studies have shown that there could be a cell receptor called ENaC responsible for salt detection in lab mice along with a second receptor which detects how salty solutions are, but the reality is that even though we put a man on the moon, and have harnessed nuclear energy to power our cities, we still don't know the specific receptor mechanisms involved in tasting chips with and without salt.

How the detection of taste becomes the perception of taste

When various foods are turned into electrical impulses by the taste buds, a combination of three nerves carries taste information away from the mouth to the brain. These three nerves send information to an area deep in the brain called the medulla. From the medulla, electrically coded taste information is transmitted to an area called the 'primary taste cortex'.

The primary taste cortex allows us to *perceive* the taste of our food, and forwards information to various other centres in the brain. One of these is the amygdala, responsible for the production and release of dopamine, the 'feelgood' chemical. It drives us to feel a sense of reward or pleasure when we eat something delicious (and explains why I still go a bit wobbly-kneed when offered lamb samosas).

Food can drive our eating behaviours in ways that we don't always realise, and, as we just learnt, there is a chemical link that explains why a mouthful of bittersweet chocolate fondant, with its oozing molten flowing centre, can trigger vivid memories of pleasure, lust and even love. But food memories work both ways; the same mechanisms can also explain why, say, having eaten one rotten oyster on holiday back in 2013 can leave you with a lifelong aversion to those tiny balls of sea snot.

Previously, scientists thought that the primary taste cortex in the brain contained specialised nerves that responded mainly to one particular taste sensation, but that these nerves had the potential to also sense other tastes at a lower level. However, studies have used MRI techniques to look at the brains of animals that are given a variety of different solutions to taste, and researchers identified that discrete areas light up in the brain when bitter, salty, sweet or umami solutions are applied to the tongue. As a result, a 'gustotopic map' of taste perception in the brain has been proposed, suggesting separate rather than diffused areas for the detection of tastes.

Although this is intriguing in principle, many in the scientific community feel that this gustotopic map is far too simple – a one-dimensional explanation for a complex, intricate medley of processes. For example, no area in the brain responsible for detecting sour tastes could be located in the experiment, raising further questions as to how it identifies the taste of sour foods. But, as with all scientific advances, questions and challenges to theories only serve to push our understanding of a subject further, and the quest to understand how taste works is no different.

Can knowing about different taste buds impact my cooking?

In my life, I have been lucky enough to meet some cooks who are truly brilliant. They manage their kitchens with an admirable dexterity and have a knack for identifying that missing ingredient that completes their dish, giving their food a certain indescribable *je ne sais quoi.*

I once had a cleaner of Indian Gujarati origin called Jaya, who also helped me prepare food in my home on the days when managing a young family and career as a doctor became an impossible balancing act. She was vegetarian and made the most glorious curries. I was particularly fond of a stir-fried white cabbage curry that she was able to put together in minutes. I noticed, to my surprise, that she added both salt and a good lump of jaggery (unrefined cane sugar) to the wilting cabbage leaves in the frying pan. I had never thought about adding sugar to any curry I had made, but it made complete sense. The sweet notes of jaggery complemented the slightly bitter-tasting cabbage leaves and the savoury seasonings that Jaya chose to make the humble vegetable irresistibly tasty and the dish so much more than the sum of its component parts.

I have thought often about what it is that makes cooks like Jaya so brilliant. Yes, they cook with heart, for pleasure and comfort, but I feel that there is more to the picture. Ultimately, instinctive cooks are able to use their palates to their advantage. Whether it is that extra squeeze of lemon juice over a sweet honey-roasted caramelised chicken or that pinch of rock salt over bittersweet melting dark chocolate, these cooks subliminally understand the components of flavour – salt, sweet, sour, bitter and umami notes – and are empowered to use them to create balanced, flavoursome dishes that arouse a variety of different taste buds in each bite.

Alas, not all of us can be like these instinctive, intuitive cooks who just effortlessly understand the workings of taste and the building blocks of flavour, prompting them to add, say, a pinch of sugar to an acidic tomato sauce or in a savoury curry, a touch of acidic balsamic vinegar with the bitter, smoky char of sweet caramelised roast vegetables, transforming them from something ordinary to a memorable dish that is hard to resist. Analysing the taste components of your food and diversifying the taste sensations you use, brings you one step closer to being successful in your kitchen.

I am certainly not the only cook who believes in the power of harnessing different taste sensations to enrich dishes. Various Far Eastern cultures have talked about the building blocks of flavour as an ideology for creating delicious food for centuries. I recall watching a documentary about Thai food as a child in which a passionate local cook was talking about using sweet, sour, spicy and salty flavours

Is there a taste bud for detecting fat?

One of my favourite flavours is the taste of fat. Butter, ghee, olive oil, full-fat cream, the crispy fat on barbecued lamb chops – these are the tastes that really set my food passions ablaze. I cannot think of anything more delightful; the stereotype of chefs adding huge amounts of butter to everything to make it taste that bit more delicious is an image that really connects to my own personal idea of food and taste.

The question of whether there are specialised taste cells responsible for perception of fat, or whether fat is more of an oral texture (or to use its technical term, mouthfeel), remains very much unanswered. Time will tell whether a specific taste bud to detect fats is identified, but it seems that some sort of oral fatty-acid sensitivity may exist that has a functional significance in humans; research has shown that people (and animals) with a decreased sensitivity to the taste of fat will eat more of it, and that this, in turn, may have the effect of increasing their weight.

I posed the question taste to my four-year-old son, asking him what different tastes he knows about. As well as sweet, salty and sour, he also mentioned 'the butter taste'. Quite a few of the children I have posed this question to in children's cookery classes seem to also identify fat as a distinct taste.

I unintentionally revisited this topic the other night after serving some exceptionally creamy, butter-laden mash on the dinner table. I made a passing comment directed at my husband about how rich it tasted and my son, who I am now convinced is destined for a career in law, announced: 'See, Mummy, I told you butter was a taste! The butter makes the mash yummy, even though you said before that it wasn't a proper taste!' I stand corrected.

like the 'notes of a musical chord' to create harmonious dishes; while one note can dominate a dish, the others should also be there in the background to some extent. Contemporary cookery authors like Samin Nosrat, who has written *Salt, Fat, Acid, Heat*, as well as Yotam Ottolenghi (*Flavour*) and Nik Sharma (*The Flavor Equation*), provide an invaluable resource for those who wish to further learn about the foundations of taste or should we say 'tasty'.

Are our senses of taste and smell related?

Taste is instinctive and involuntary; we do it multiple times a day without really thinking about it consciously. We don't choose to turn on our taste buds when we eat, and we have no way of not tasting what is in our mouths.

So far, I have explained a little about how we taste salty, sweet, sour, bitter and umami, and I also hope you have enjoyed me sharing some of the things that make the world of taste so weird and wonderful. But now, let's change focus slightly and look at how our perception of taste relates to another very important food-related sense: our sense of smell.

At university, I had a friend called Rose who had grown up in the countryside before starting university in south London. She had always suffered with allergies and hay fever, and often had nasty bouts of sinusitis. Upon arrival in London, things went from bad to worse, and for the entire first year Rose suffered with sinusitis constantly. Gradually, these issues wore away at her enjoyment of food until one day, while eating jacket potatoes with tuna in our medical-school canteen, she mentioned to me that she thought she had almost totally lost her sense of taste. The jacket potato and tuna tasted of cardboard, she professed, clearly quite upset by the situation.

'I can't really even taste curry,' she sighed.

We talked about her symptoms at length, trying to figure out why this had happened. Being so obsessed with food, a life without the pleasures of taste, to me, seemed intolerable, almost unthinkable. So, as you would expect from a couple of medical students, we took it upon ourselves to work out how we could change this state of affairs. Maybe she had taken too many antibiotics that altered her taste perception? Or perhaps she had too much mucus coating her tongue?

After some back and forth, I asked Rose whether she could still smell her food, and following a few moments of deliberation she conceded that her sense of smell had vanished almost completely. Could this be the reason for Rose's malfunctioning taste buds? Could it be that taste and smell were somehow linked so intrinsically that losing her sense of smell was also robbing Rose of her sense of taste?

When Rose eventually managed to visit an ear, nose and throat specialist, it transpired that she had developed a polyp (a small, non-cancerous growth) at the back of her nose, which was blocking her sinuses, resulting in the loss of her sense of smell. When the polyp was finally removed a full six months later, virtually overnight Rose's sense of smell returned, and with it her sense of taste. To celebrate, we went to the place most full of flavours that we could think of: our local Bermondsey curry house. We toasted the return of flavour into Rose's life, and all the accompanying joy that only a good curry can deliver.

How do our senses of taste and smell interact?

So, clearly, olfaction (or the sense of smell) contributes heavily to our sense of taste. But to what extent? Actually, it looks like the answer to that is *a lot*. Some researchers claim that up to 80 per cent of what we experience when biting into our food does not even come from our taste buds, but rather from our sense of smell.

Up until recently, scientific literature had, on the whole, agreed that human beings were able to discriminate around 10,000 different odours. To me, this seemed like quite a lot. I mean, I couldn't name 10,000 distinct smells for you right now. But when experiments were designed to challenge this figure and examine the resolution of human smell more accurately, researchers were faced with some intriguing results.

Scientists estimate that humans may be able to identify a trillion different olfactory stimuli – a far cry from the previous estimate of 10,000. (To put this astonishing number in context, humans can see around 8–10 million different colours and maybe half a million tones.)

In fact, the olfactory system outperforms virtually all of our senses when it comes to the number of different stimuli it can discriminate. Humans have most likely only developed a predominance for relying on sight simply because we are able to process visual input around ten times faster than smell input. It's also more sensible to train your body to react to the sight of a tiger than the smell of one.

But back to the link between smell and taste. The theory is that each morsel of food contains thousands, if not millions, of what are known as volatile aromatic compounds. A volatile compound is essentially a chemical that moves through the air, and aromatic here refers to any chemical that the human olfactory system is able to smell.

A volatile aromatic compound emitted from food on our plates will be drawn to the back of the nose and throat through our breathing, and bind with a specific receptor cell which, through a complex network of nerves, will fire a signal to the brain where the aroma associated with that compound can be interpreted. When that food enters the mouth and is broken down by chewing, further aromatic molecules will burst out, hit the back of the throat and nose and, again, allow us to experience those magical complexities of flavour.

In essence, volatile compounds help create flavour profiles, so when you describe something as fruity, floral, citrussy, herbal, earthy or meaty, it is your interaction with them at work. And while the exact rules governing *how* olfactory and taste-bud information is integrated are not fully understood at this moment in time, what we do know is that they *are* integrated and that, as Rose demonstrated, our experience of food depends heavily on both senses working together in harmony. The most cutting edge science suggests the same aromatic compound, sensed via the normal sniff versus being sensed from within the mouth, can evoke slightly different perceptions and activate different parts of the brain.

So, the next time you have a roast chicken in the oven and your partner comes home and tells you dinner smells amazing, you can sit them down and tell them about the volatile aromatic compounds from the roasting chicken, and how they interact with their olfactory system, resulting in the experience of smell. Maybe get a whiteboard, draw some diagrams. They'll love it.

I suppose instinctively, this all makes sense. When you have a blocked nose because of a nasty cold, your sense of taste is diminished because those volatile aromatic molecules become unable to fight through the sticky layer of snot in order to reach the back of your nose and throat. I have met patients with particular head injuries who have lost their sense of smell, because the delicate nerves carrying sensory information to the back of the brain have been damaged or torn. I have met others undergoing radiotherapy due to head or neck cancers who often complain of the flavours of their favourite dishes not being quite the same as before. One such patient told me that unless she added extra salt, sugar and chilli to compensate, her food tasted no more flavoursome than a wad of paper. It also explains why Rose's sense of smell and taste returned to normal so rapidly after her nasal polyp was removed.

If you still don't believe me when I talk about how vital the sense of smell is to influencing taste perception, here is an interesting experiment (in the interest of science, obviously). Try closing your eyes and pinching your nose, and then put a jelly bean in your mouth and give it a good chew. My guess is that you won't be able to identify exactly which flavour it is, even if it may seem a little sweet or even sour at first (because the taste buds are obviously still working, just without the amplifying effect of their partner sense). Now gently un-pinch your nose and allow all those volatile aromatic compounds to float to the back of the tongue, and up the back of the nose. Has anything changed? I imagine that the distinct flavour of the jelly bean, as well as the underlying sweet or sour tones are much more pronounced than when you were tasting them with your nose pinched.

Understanding the sensory component of olfaction and its relationship to taste is essential for a chef because, I am certain, it helps you to cook more refined, more nuanced dishes. Thinking back to my time on *MasterChef*, I recall I was once cooking a dry spiced lentil dish when judge and professional chef John Torode walked past and casually commented that something was about to burn. I was a little surprised as I wasn't expecting any dish to be complete for at least a few minutes, and burnt food is the ultimate irrecoverable culinary disaster.

However, when I cast an eye upon my lentils it transpired that I had accidentally left the heat on high, and indeed, if I had left them for another minute or so, they would have started burning. John, ever the teacher, looked at me. 'Saliha,' he said, 'don't forget to use your sense of smell. It'll help you recognise when your dishes are prepared to perfection.' And off he strolled.

Now, to many of you this may seem like sensible, but not groundbreaking advice, and you'd be right. But the practical implications in my kitchen have been almost immeasurable. You can train your nose to distinguish a wider range of smells, becoming what professional sniffers (yes, they exist) call 'odour aware'. It's a particularly useful skill in the kitchen, where smells can tell you whether a food is raw, perfectly cooked or burned beyond recognition. For example, I can now rely on my nose to tell me when to flip my pancakes. Where once I might have flipped prematurely, I now manage to sniff out, with a very respectable rate of success, the perfect golden underside before the toss. Similarly, after a lot of practice, I now often know when a roast dinner or pasta bake is ready by trusting the messages from my nostrils, and not just my eyes. Even cooking rice has become easier since teaching myself to recognise the smell when it is cooked perfectly. As a consequence, my timer is now languishing at the back of a kitchen cupboard somewhere.

This has increased that intuitive connection that I feel with the food I cook, which has greatly contributed to my sense of gastronomic happiness and confidence in the kitchen. This skill may be something you want to add to, or improve in your cooking practice. If so, I encourage you to trust your nose because it, more than any of your other senses, can tell the difference between when your food is good and when it is great.

Inherited hatred (or, The Story of Coriander)

The mysterious case of *Coriandrum sativum*, coriander (or cilantro, to our counterparts in the United States), is an odd one. People fall definitively into either the love or hate camp, but why? Personally, I love the citrus-tinged, bucolic appeal of coriander, and I use it generously in my recipes. I grew up eating curries that were sprinkled liberally with freshly chopped coriander and I always felt it brought them to life. But, if you are a fervent opponent of this herb, the experience of ingesting it can be utterly detestable, one that is commonly likened to swallowing a bar of soap.

A sizeable proportion of my friends, all of whom I consider to be of relatively sound mind and body, fall into the latter camp. When Julia Child, queen of American cookery, was asked by broadcaster Larry King about the foods she hated, she cited a hatred of 'cilantro' so deep that it tasted 'dead' to her, and if it appeared in front of her she would pick it out of her food and throw it on the floor. And if you think that is extreme, I urge you to put this book down, go make a cup of tea and read some of the comments on the 'I F***ing Hate Coriander' Facebook group (or the 'I Hate Cilantro' group if you're American).

There is growing evidence to suggest that a dislike for coriander is actually coded within our DNA. Observations of twins showed that while around 50 per cent of non-identical twins held the same opinion of it (which is the 50/50 spread that you would expect), a full 80 per cent of identical twins (who have the same DNA) shared the same opinion as their twin about either their love or hate of the herb. It may not sound like much, but a 30 per cent difference in the two groups is a very significant jump in scientific terms, and could point to some sort of inherited characteristics that shape a person's love or hate relationship with certain foods.

As ever, the story is nowhere near clear. The results shown in the twin study above can definitely be interpreted as an inherited genetic trait, but the full picture is still blurry. Scientists will continue to add in details as research progresses. And while nurture is also important (i.e. the way the subjects are brought up and the messaging they hear about particular foods from their parents and peers), the evidence suggests that inherited food preferences may be a real thing.

Coriander isn't the only food involving what could be an inherited genetic preference. The humble Brussels sprout, broccoli, garlic, a variety of salad leaves and spices also fall into the same gene-based love/hate conundrum. People feel that they taste, well, bitter. For the record, I love them all, and I will go to my grave arguing that Brussels sprouts are not just for Christmas.

I have designed a number of recipes that celebrate these divisive ingredients and they serve as fantastic starting points for those of you who are attempting to convince a loved other of their culinary merits. How about trying a charred garlic and coriander dressing? Or perhaps you will be inspired by a coriander and walnut 'zhoug' to douse your roast lamb. Deep-fried broccoli florets in pakora batter, till feather-light and crisp, might catch your fancy. Or maybe roasting some sprouts in a searing hot oven and serving them with a curry-tinged dipping sauce will change your opinion of sprouts for ever.

Try and put any prejudice aside and see what you really think. I for one can tell you that, using these recipes, I have convinced a few coriander-, broccoli-, sprout- and garlic-phobic friends that their deep-seated hatred was ill-founded using these recipes, so they are certainly more than worth a try. If, at the end of it, your dislike for these ingredients persists, and the bitter taste remains intolerable,

I suppose you can accept that it was just not meant to be, nature having prevailed over nurture.

Out of all of the flavours, it is the bitter tastes that have most captivated my interest. Bitter stimuli are extremely diverse in their chemical structure. The function of being able to sense them seems to be related to our capacity to detect poisonous substances, and as vertebrates rose from the ocean and on to land more than 500 million years ago, the ability to taste became a key part of our primitive survival mechanism. Proof of how important it was for our ancestors to avoid poisons can be found in the fact that the different bitter receptors far outnumber sweet and other taste receptors on our tongues. It's also why bitter tastes can overwhelm us much more easily than sweet or sour.

Neurological and behavioural studies in humans, as well as rats and monkeys, show that these species can actually distinguish between a range of different bitter stimuli. There are suggestions that the nerve fibres responsible for detecting specific types of bitterness may even be grouped in discrete bundles in the brain. The next challenge for the culinary world is to come up with names for the various types of bitter tastes that exist!

There is also evidence which seems to show that sensitivity to bitter tastes might be a genetic trait. There is a solution called phenylthiocarbamide, or PTC, which around seven out of ten of us would find tasteless, while the remaining three, upon ingesting it, would experience the most repugnant and intolerably bitter taste imaginable. These three out of ten people are essentially 'bitter supertasters' and are less likely to enjoy cabbage, broccoli, Brussels sprouts, beer, coffee and other such pleasures throughout their lives.

To those three out of ten people, my condolences go out to you.

Melon, Mango and Cucumber with a Coriander and Burnt Garlic Dressing

Serves 2

Part of the appeal of coriander lies in the fact that its lemony, citrus-tinted, grassy notes make it the perfect pairing for a host of different savoury, sweet and fruity flavours. Those of you who perceive coriander as tasting soapy will be pleasantly surprised by how appealing this combination of ingredients really is. Burning garlic on the cooker is a Bangladeshi tradition, imparting a smoky back note to the final dish.

Ingredients

125g cantaloupe melon, diced into 1cm cubes
125g mango, diced into 1cm cubes
125g cucumber, deseeded and diced into 1cm cubes
15g coriander, finely chopped
2 garlic cloves
1 tablespoon maple syrup
Zest and juice of 1 lime
1 red chilli, finely sliced
2 teaspoons tamarind chutney
1 tablespoon olive oil
Salt to taste

Method

1. Start by mixing the melon, mango, cucumber and coriander in a bowl.
2. Place the garlic cloves on the end of a fork and carefully burn them in the naked flame of your cooker (or in a dry non-stick pan on an electric hob) until they are fully charred and black on all aspects – this takes around 3 minutes.
3. Crush the burnt garlic to a pulp with the side of your knife and add it to the bowl, along with the maple syrup, lime zest and juice, red chilli, tamarind chutney, olive oil and a sprinkling of salt. Give everything a good mix before serving.

Note: it is perfectly acceptable for the fruit to be cut into chunks larger than 1cm, if you don't have the patience to chop so finely!

Broccoli Pakoras

Serves 4

If someone dislikes broccoli, you can near enough guarantee that the best way to convince them that their hatred is ill-founded is to serve them the vegetable deep-fried in a crispy batter. On a cold and drizzly afternoon, these crisp pakoras with a side of tamarind chutney and the greedy fingers of family members rushing to grab the fresh batch out of the fryer is my idea of broccoli heaven.

Ingredients

75g gram flour
1 heaped tablespoon cornflour or rice flour
½ teaspoon bicarbonate of soda
1 teaspoon chilli powder
1 teaspoon cumin seeds, coarsely ground
1 teaspoon coriander seeds, coarsely ground
1 teaspoon nigella seeds
½ teaspoon salt
2 tablespoons lemon juice
200g tenderstem broccoli, roughly chopped into 3–4cm pieces
Vegetable oil, for deep-frying

To serve:
½ teaspoon chaat masala (optional)
Tamarind chutney

Method

1. Mix the gram flour, cornflour, bicarbonate of soda, chilli powder, ground cumin and coriander, nigella seeds and salt together in a bowl. Whisk in the lemon juice and just enough water to form a batter that has the consistency of double cream. Mix the broccoli into the batter and stir to coat well.

2. Heat the vegetable oil in a large pan or deep-fat fryer until a drop of batter sizzles in the oil. Deep-fry the broccoli pieces in small batches until they turn crisp and the batter takes on a deep golden colour. Remove and drain on kitchen paper while you make the next batch. This takes around 4–5 minutes.
3. Serve with a final sprinkling of chaat masala, if desired, and a side of tamarind chutney.

Note: if broccoli isn't enough to contend with, you could serve these pakoras alongside a coriander green chutney. Just blitz together 125g raw cashews with 125g golden sultanas, 60g chopped coriander, 75ml lemon juice, a green chilli and a sprinkling of salt, till smooth. Just don't tell the diners it is made from coriander, and test their reaction!

Roasted Brussels Sprouts and Curried Yoghurt Dip

Serves 4

The struggle that Brussels sprouts face is not that people dislike how they inherently taste, but that they are so often cooked extremely poorly. Boiling a Brussels sprout until it's mushy is, in my view, a cardinal sin. Culinary justice lies in either roasting them in a hot oven or shredding them and stir-frying rapidly.

Ingredients

500g Brussels sprouts, halved lengthways
2 tablespoons extra virgin olive oil
1 tablespoon honey
1 teaspoon turmeric
½ lemon
1 tablespoon mayonnaise
125ml yoghurt
2 tablespoons vegetable oil
10 curry leaves
1 teaspoon mustard seeds
Seeds of 1 cardamom pod
1 green chilli, finely chopped
Salt, to taste

Method

1. Preheat the oven to 200°C fan. Toss the sprouts with the olive oil, honey, and ½ a teaspoon of the turmeric and season with salt. Tip them on to a baking tray and roast in the oven for 10–15 minutes. Remove from the oven and grate over the zest of half a lemon.

2. Mix together the mayonnaise and yoghurt in a bowl. Heat the vegetable oil in a non-stick pan till it is hot, but not smoking, Add the curry leaves, mustard seeds, cardamom seeds, remaining turmeric and green chilli to the hot oil, stirring quickly to ensure nothing catches. When the spices start popping and the curry leaves are crisp, remove the pan from the heat. Use a slotted spoon to remove the curry leaves and set them aside.

3. Pour the hot oil on to the yoghurt-mayo mixture and mix it through, along with the juice of half a lemon and salt to taste. Work carefully, standing back from the cooker, as the spices can spit out of the pan and burn.

4. To serve, scatter the crisp curry leaves over the roasted sprouts and yoghurt dip.

Garlic Roast Leg of Lamb with Coriander Walnut Zhoug

Serves 6

Zhoug is a bit like the Middle Eastern version of salsa verde. It is rather delightful as an accompaniment to lamb, but is also fantastically versatile, finding a place in salads, soups, roasted vegetables or even swirled into hummus. What I dearly love about this dish is that with minimum effort, the results are quite astonishingly good. People think you have slaved away for hours at a stove, but in reality, all it takes is a mortar and pestle, a slab of meat and an oven. Just the type of culinary deception I love. I have been deliberately heavy-handed with the garlic; I have a long-standing love affair with the bulb, but you can tone it down if you wish.

Ingredients

For the lamb:

4 tablespoons olive oil
All the cloves from 2 whole garlic bulbs, crushed in a mortar and pestle
2 teaspoons red chilli flakes
1 heaped teaspoon oregano
2 tablespoons sherry vinegar
1.5kg leg of lamb, deboned and butterflied
Salt, to taste

For the walnut zhoug:

2 garlic cloves
1 teaspoon toasted cumin seeds
½ teaspoon ground cardamom
100g toasted walnuts
40g coriander, roughly chopped
3 fresh jalapeño chillies, finely sliced
4–5 tablespoons olive oil
1 tablespoon sherry vinegar or 2 tablespoons lemon juice
Salt, to taste

Method

1. To make the lamb, prepare the marinade by mixing together the olive oil, garlic, red chilli flakes, oregano and sherry vinegar. Stab the lamb around twenty times with a skewer; the perforations will help the marinade seep through the meat. Ensure that the lamb is more or less uniform in thickness, and if it is not, then make some incisions in the muscle belly to flatten it out. Rub the marinade all over the lamb and refrigerate, ideally overnight, but for at least 3 hours before cooking.

2. Preheat the oven to 180°C fan. When you are ready to cook the lamb, remove it from the fridge for at least 30 minutes to bring it back to room temperature before placing in the oven. Season liberally with salt and place the lamb in the oven on a

deep roasting tray along with all the marinade. The lamb needs to cook for 20 minutes if you want it rare, for 30 minutes if you like it medium and 40 if you like it well done. I tend to opt for medium. When you remove the lamb from the oven, allow it to rest covered with some foil for 20 minutes before transferring to a platter and carving into thin slices, cutting with, rather than against, the grain.

3. Prepare the zhoug while the lamb is in the oven. Start by placing the garlic in a mortar and pestle, along with the cumin seeds and cardamom. (You can use a food processor if you wish, but I must admit that results are better when using an old-fashioned mortar and pestle coupled with some elbow grease.) Pound until the garlic cloves have become paste-like. Now add the walnuts. The idea is to pound them enough for them to break down into a nutty rubble, but not lose their texture completely. To complete the zhoug add the coriander, jalapeños, olive oil, vinegar and a generous amount of salt. Mix, using the pestle to combine everything together. Taste and adjust the saltiness or acidity to your desire.
4. There will be some sticky burnt bits and juices at the bottom of the roasting tray. Make these into gravy by placing the roasting tray on your cooker over a medium heat and adding around 250ml water. Scrape all the sticky meat juices from the bottom of the pan with a wooden spoon. After 3–4 minutes, you will have a concentrated gravy. Spoon some of this over the slices of lamb – try not to use all of it, as it will be quite intensely salty and garlicky.
5. To complete the dish, spoon the walnut zhoug liberally over the lamb slices and serve alongside crusty bread or boiled new potatoes.

Eating on board Flight PK757

Most people I know hate the taste of airline food. 'There is no f***ing way I would eat on planes,' Gordon Ramsay has said. Having worked with airlines in the past, Ramsay has often complained that the way that food is prepared, stored and arrives on board, along with the dry convection that is used to reheat it, renders it virtually unpalatable. Many of the chefs I know opt to stay hungry, or to eat pre-bought food to satiate their in-flight hunger pangs.

Now, the opinion that I am about to share is not widely held, but while most aeroplane meals have generally left me feeling a bit nauseous, I have also experienced in-flight meals that I would wholeheartedly describe as delicious, on par with many ground-based kitchens.

PIA stands for Pakistan International Airlines, an airline which in 2019 didn't even break into the Skytrax top one hundred in the world. For years, PIA has run flights from my hometown of London to my grandparents' hometown of Lahore, and when I was a child, PIA served nothing but traditional Pakistani food on board. What I mean by that is: if you fly PIA, you had better like curry.

One particular year, I peeled back the silver foil lid of my in-flight meal to reveal a rectangular plastic plate full to the brim with lamb biryani. The scent of pungent spices hit my nose through the pressurised air of the cabin, filling my nostrils with the unmistakable anticipation of deliciousness. The rice grains were perfectly cooked, having somehow resisted the variations in temperature that usually come about from the dry convection treatment. Seasonings were generous, including healthy doses of ginger and garlic, finished off with a combination of finely chopped red and green chillies for that extra kick. The spice level, while high, was not a problem for a Pakistani palate that was used to higher than average levels of heat.

Sitting with a numb bottom, in an uncomfortable seat with no leg room, the fatigue and stress of the journey was temporarily forgotten and I finished off the entire meal, while fighting my little sister for the tiny amount of elbow room. We ate quickly and in silence, until my mother, finishing the last of her own biryani, said: 'Well, that wasn't bad at all, was it? Looks like the holiday has already started!' She then took a sip of her tomato juice – an odd choice, I thought at the time, given that she never, to my knowledge, drank it at home.

In retrospect, I have often wondered why airline food in general tastes quite so awful and mundane, and why this particular meal was so scrumptious. And it was only when I discovered the work of Oxford scientist Professor Charles Spence that I came to understand the effect that flying has on our taste buds.

There are three key reasons why the taste of airline food is usually so lacklustre, according to Spence. Firstly, the pressurised cabin in an aircraft alters our sense of smell and taste by drying out the nasal passages. Moisture is very important for the transmission of scents through the air (for example, the smells that surround you after a heavy rainfall are actually always in the air, but the extra moisture that the rain brings allows your nose to temporarily detect a much richer range of odours). Secondly, low levels of moisture in the air can reduce the sensitivity of taste buds by up to 30 per cent. And finally, that extra background noise during a flight actually has an effect on more than your ability to hear the movie you are watching – it also impedes your perception of taste.

Having said that, it seems that sour, umami and bitter tastes are less affected on aircraft than sweet and salty flavours, so taking a packet of sour sweets into the air with you for a cheeky snack is probably a safe bet. My mother opting for tomato juice on flights also now makes sense, because it is a drink dominated by a rich umami taste. And while enjoying her tomato juice, she would always add an extra sachet of salt to her glass. Without consciously realising it, she was, it seems, compensating for the dampened salt taste at high altitude.

When it comes to spices, there is evidence that curry and lemon-grass actually become more intense in the sky, while cinnamon, ginger and garlic tend to maintain their taste. This explains, at least in part, why that biryani tasted so good all those years ago. Jude Law has the right idea; he is reported to always take Tabasco sauce on board his flights in an attempt to make food more palatable. There are also reports of astronauts requesting hot sauce to improve taste quality of their food. To these astronauts I would say, why not try a PIA biryani on board instead? I am confident they would not be disappointed.

So where are we in our journey towards digestive health and happiness? Well, perhaps this chapter has helped you to realise that the journey starts when your food is tasted, properly. I believe that many of us take the experience of 'taste' for granted; eating can be habitual just like other activities of daily living, washing, dressing, cleaning, etc. I hope you are now encouraged to actively taste your food and engage with the complex multi-dimensional experience that we grossly simplify and call 'taste'. Spending some time contemplating your wonderful (and not so wonderful) taste experiences and engaging with the feelings of both pleasure and revulsion that they elicit is liberating, and one of the foundations to fostering a fulfilling relationship with food.

In the next chapter, we will see how our experience of food is linked not only to taste and smell, but also to texture. We will explore how different food textures can create different experiences in a bite which, used effectively in the design of a dish, can elevate a good plate into a fantastic one, separating the good chefs from the truly great ones.

Summary

- The primary tastes detected by the taste buds are sweet, salty, sour, bitter and umami.
- Taste and emotion are inexorably linked, and food has the ability to elicit both pleasure and revulsion. We are still not entirely clear about how tastes are mapped in the brain, but your food memories do have a huge impact on what you choose to eat in certain scenarios.
- In time, we may find out if there is a specific taste bud to detect fat. It seems that we may possess a certain 'fatty acid sensitivity', whereby those of us with a lower sensitivity end up eating more fat than others. Consider whether you have high or low levels of fat sensitivity compared to others around you.
- Our sense of smell contributes hugely to the creation of flavour profiles. You can use this to your advantage. A greater awareness of the sense of smell in your kitchen will help you to become a better cook.
- If there are certain foods you do not enjoy, particularly bitter-tasting foods like coriander, broccoli, sprouts and garlic, it could be that your dislike is genetic. So, don't beat yourself up too much if you don't like sprouts.
- Next time you travel in an aircraft, opt for tomato juice and a heavily spiced, strong-tasting meal. It will be far more enjoyable than the bland food everybody else is eating.

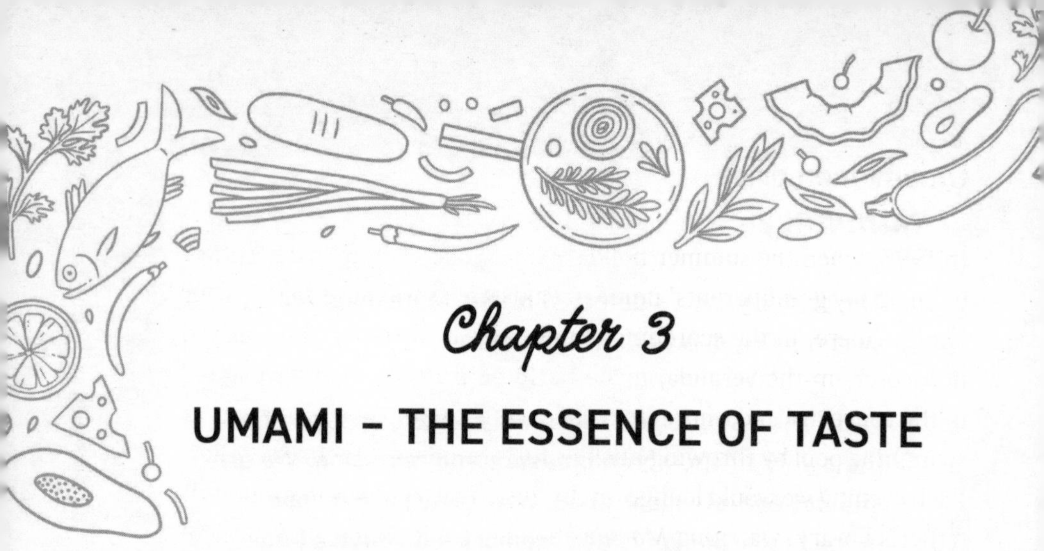

Chapter 3

UMAMI – THE ESSENCE OF TASTE

As anyone who has a favourite food will tell you, certain tastes can affect your relationship with food in profound ways, while others maybe don't resonate with you as much for any number of reasons. For some people, it's a sour lemonade that they enjoyed on summer days in their youth. For others, it's the sweetness of a birthday cake that affects their sense of memory and stirs a warm recollection from deep within their brains. For me, I have learned that the taste that I connect with the most is one that until recently didn't even have a name. Welcome to umami.

Before sitting down to write this chapter, I crept quietly into my kitchen, toasted a crumpet and covered it with butter and a generous helping of earthy, umami-rich Marmite. I wasn't even hungry, to be honest, but Marmite on crumpets is one of those snacks that doesn't require hunger before the cravings kick in. Marmite, the umami-rich yeast extract that we either love or hate, has, for me, the most satisfying, lingering, savoury, meaty flavour. It is therefore quite fitting that my chapter on umami, the mystery fifth taste created by the detection of glutamate and related amino acids, began with a bite of it.

Umami memories

In 1995, when the summer holidays finally arrived, my parents flew us out to my grandparents' home in Pakistan, as was customary in our family. There, in the scorching summer heat, my sister and I played hopscotch on the veranda, made perfume from the flowers planted in the courtyard and, more than once, converted the bathroom into a swimming pool by throwing buckets full of cold water at one another. Each evening we would lounge on the dusty maroon sofas in my grandfather's library, watching *Matilda* on video while eating cold cubes of honey-sweet, fleshy mangoes. Now, people throw the word 'idyllic' around a lot these days, but I'll be honest, this wasn't far off.

But even when you are given all the freedoms in the world, there are rules that must be followed. One of these was that in the blistering heat of the Pakistani summer, entry to the family kitchen was barred to all but the *khansama*, or chef. His name was Sharif, and the kitchen was his domain.

It was my grandmother's rule. In her Pakistani-grandparent kind of way she would often voice her concerns about our fading complexions, and how we would wilt if we spent even a second in the blazing heat of Sharif's kitchen. While reasonable on one level, on another level, if you forbid a child to do something, there is a 100 per cent chance that they will do it at the first opportunity that presents itself. And, in the words of Matilda, the famous child who was forbidden to do things but did them anyway, 'Never do anything by halves if you want to get away with it'.

And so, when it was time for my grandmother's afternoon siesta, I crept away from the comfort of our air-conditioned bedroom and into the bowels of the kitchen. As I mentioned, the kitchen was very much Sharif's territory. It was organised by him in a way that only he understood, incredibly old, but generally clean. There were jars full of spices, pickles and potions on rickety-looking shelves that were covered in old pieces of newspaper bearing headlines about the moon landing and the

Vietnam war. Sharif cooked delicious meals each night, and had been doing so for our family for over thirty years. Although his repertoire was mostly that of regional Pakistani cuisine, every once in a while, he would venture out of his comfort zone and dip into what he would call 'food of the outside world'.

One such globally inspired dish was Keema Spaghetti (see p. 73), a Pakistani take on spaghetti Bolognese. Now, the food purists among you may judge me for this, and I get where you are coming from. However, I would encourage you to reserve judgement until you have tried Sharif's Keema Spaghetti for yourself. It is utterly addictive, and probably the dish I loved most during those Pakistani summers. Chunky strands of thick spaghetti were covered with slow-cooked fatty lamb mince, sweet peas, carrots, chunks of fried potato and generous amounts of garlic. The dish, as a whole, was not particularly saucy or oily, and the layer after layer of savoury richness that filled every mouthful was an experience not easily forgotten. Obviously, he served it with a side of tomato ketchup.

To my surprise, Sharif welcomed me into his kitchen, though with a mixture of caution and enthusiasm. But armed with a notebook and razor-sharp pencil, eight-year-old me was ready to take risks in order to discover the recipe for Keema Spaghetti. In my mind, I was doing my mother a favour by learning it, because this was a dish that *had* to be introduced into our lives in the UK. Waiting a whole year to eat it again was not an option.

Sharif sat me on a tall stool beside him and talked me through how he fried off the onions, balanced the concoction of pungent and temperate spices until it was just right, and how he browned the mince until it was a deep, dark golden colour. His control of the knife, the heat of the pan and the natural order he created in an otherwise chaotic kitchen mesmerised my young mind. He spoke neither too much nor too little, giving away little emotion, but I knew he was enjoying delivering his tutorial just as much as I was enjoying attending it.

I asked Sharif if I could give the contents of the pot a good mix myself, but my request was, of course, immediately denied. After all, even being outside the air-conditioned room at this hour was already perilous; he wasn't about to risk me spilling hot keema all over my arm. But still, with his long, greying beard, slender physique and gold-rimmed glasses that magnified his eyes until they took up most of his face, he toiled away while I watched, pots simmering, the scent of spices and meat filling the air around us.

In around forty minutes, the symphony of Keema Spaghetti was, it seemed to me, complete. But as the ribbons of pasta and mince were about to meet one another in holy matrimony, Sharif paused. 'One last secret,' he said. He leaned towards me, and from a black carrier bag he produced a small clear plastic bag filled with translucent white powder, similar in consistency to coarse salt. As he turned the packet over and handed it to me, I read the bright red label: 'AJI-NO-MOTO seasoning', printed in large, stylised lettering.

I let out a loud squeal and started laughing hysterically. You see, if you translate *ajinomoto* into Urdu, it means 'fat ghost', and to an eight-year-old, fat-ghost seasoning was unmistakably hilarious. 'It is a special type of China salt you use instead of normal salt – makes the dish taste really, really good,' Sharif explained. 'The essence of deliciousness.'

The essence of deliciousness? His words felt to me like a secret, a gift of knowledge for proving my worth by sitting nicely in his kitchen. 'Ok, I believe you!' I proclaimed. 'Fat Chinese ghosts must taste salty!' I joked, still giggling at the name but happy that he had trusted me with his secret ingredient.

Once my laughter had subsided, I was offered a fresh portion of Keema Spaghetti. It was so satisfying, each bite compelling, harmonious and comforting, like being wrapped in a cosy blanket. The type of delicious that is so much more than the sum of its component parts, complex without being overpowering.

Suddenly, from a distance I heard my

grandmother's voice booming from across the house. 'Where is Saliha?' she yelled at the top of her lungs. I ran to the dining room adjacent to the kitchen, my half-full plate of spaghetti in one hand and my recipe notes and pencil in the other. Jumping into the seat closest to me and shoving the notes under my bottom, I tried to control my racing heartbeat, act as natural as possible for a child about to be scolded by her grandmother and eat my spaghetti as innocently as I could.

'There you are, Saliha!' said my grandmother, her face stern but her voice betraying her fondness and love. 'What are you doing sitting here in the dining room, without the fan on? Look how red your cheeks are!'

I had pulled off the Keema Spaghetti recipe heist; and I had gotten away with it too.

When I was in my early twenties, my mother broke the news to me that Sharif had passed away. I felt deeply saddened but did not cry when I heard. Instead, I saved my tears until the next time I cooked Keema Spaghetti for my own family. Sharif was part of our family, and his place in my childhood and my kitchen will never be forgotten.

And for the record, 'Aji-no-moto', as it turns out, does not mean fat ghost. It is the commercial name for monosodium glutamate (MSG), the white crystalline powder responsible for the flavour we now know as umami, and it actually means 'essence of taste' in Japanese. In that kitchen in Lahore all those years ago, Sharif, without being aware of it, was introducing me to the importance of umami, and the indescribable quality that it can add to our food.

In the previous chapter, we touched on how umami is perceived at the level of the taste buds. Now we will try to explore how we can harness its power, whether from Sharif's 'China salt' or from the many umami-rich foods that exist naturally, in our own kitchens.

Sharif's Keema Spaghetti

Serves 4 (generously)

Ingredients

1 tablespoon olive oil
1 white onion, finely diced
500g lamb mince, not too lean
1 tablespoon minced garlic
1 teaspoon dried oregano
½ teaspoon turmeric
1 heaped teaspoon garam masala
1 teaspoon crushed cumin seeds
1 teaspoon red chilli powder
1 teaspoon hot paprika
3 ripe tomatoes, diced
1 beef stock cube, dissolved in 500ml boiling water
200g carrots, diced
200g frozen peas, defrosted
350g dried spaghetti
250g potatoes (skin on), diced
Vegetable oil, to fry the potatoes
Salt, to taste
1 tablespoon grated Parmesan (optional)
Bottle of tomato ketchup (mandatory)

Method

1. Place a large non-stick saucepan over a medium heat. Add the olive oil to the pan, followed by the onion, allowing the onion to soften and take on a gentle golden hue. Add the lamb mince and, using a wooden spoon, break it into smaller chunks. Crank up the heat to the highest setting and brown the mince off. You want it to release its own fat and fry off rather than stew in its own juices. Resist the temptation to keep stirring; the mince needs time to brown on each side and keeping the contents of the saucepan moving will prevent this from happening.

2. When the mince looks sufficiently brown, turn the heat down to medium and add the garlic, oregano, turmeric, garam masala, cumin seeds, chilli powder and paprika. You want the spices

to release their aroma but not burn, which is why turning the heat down a bit is necessary. Add the tomatoes, followed by the beef stock. Allow the mixture to simmer away over a medium-low heat for around 30–45 minutes, or until most of the water has evaporated and a rich, fatty, slightly moist, spiced mince remains. Taste the mixture and add salt; the stock cube you added previously is full of umami notes (or should I say commercial MSG) and will season the mince, so be cautious of adding too much salt at this stage.

3. Add the carrots and peas to the mince and allow them to cook through for a further 5 minutes. The idea is that they retain their colour but lose their crunch and soften slightly. Be wary of the mixture becoming too dry and catching at this stage. Just add a few splashes of water from the kettle if things in the saucepan start looking too dry. Once prepared, remove the mince from the heat and set aside.

4. Boil the spaghetti in heavily salted water and according to manufacturer's instructions, to achieve an 'al dente' consistency. This usually takes around 6–8 minutes. Drain the pasta, keeping a little aside.

5. Shallow fry the potatoes in vegetable oil until the potato chunks are nice and golden. This takes approximately 5–7 minutes. Drain on kitchen paper.

6. Assemble the dish by tossing the pasta into the minced lamb with about half a cupful of pasta water to moisten everything. Stir really well to combine. You want every bit of the spaghetti to be coated. Tip the tumbling cascades of spiced spaghetti on to a large serving platter and top with the chunks of fried potato and, if you wish, some grated Parmesan. Serve with ketchup (this condiment is mandatory).

Note: please do feel free to add your own favourite umami-rich condiments to this recipe – a dash of Worcestershire sauce, a teaspoon of miso paste or even a teaspoon of Marmite will add a new flavour dimension.

An umami awakening

It was about ten years ago when I first saw Nigella Lawson prepare Marmite and butter spaghetti on television. It's a dish that has gone on to become a favourite of mine, particularly when stocks in the fridge are depleted. I knew at the time that it was a divisive dish, but I was, and remain, an addict. Given that Marmite contains a massive 1,750mg of glutamate per 100g, it is the most glutamate-rich umami condiment sitting in my (or, most likely anybody's) larder.

The uses of Marmite in savoury foods are endless. I use it often to season my mince and add depth to hearty stews and slow-cooked lentils. Predictably, I adore Twiglets, the wheaty, crunchy snack that is slathered in yeast extract. I even add a hefty amount of Marmite to my Cheddar, Marmite and Chilli Cornbread (see p. 88), giving it that extra-wholesome savoury oomph. The slices of warm cornbread fresh from the oven are made all the more wonderful when smeared with softened butter and, you guessed it, yet more Marmite!

While Marmite is quite distinctive, the umami flavour in general is rather mysterious in the sense that it is difficult to describe its exact properties. Most attempts at articulating the taste of umami – words like 'delicious', 'meaty' or 'rich' – never seem to quite hit the nail on the head. Chefs talk about umami as a way of making a dish more compelling, of harnessing the natural deliciousness of ingredients or of adding savoury depth and complexity. What is interesting, though, is that even without a proper definition, many of us instinctively know exactly what flavour is being alluded to when we talk about umami.

I first came across the term umami as a contestant on *MasterChef 2017*, working in the kitchen of Michelin-star chef Sat Bains. We had to cook his dishes in the semi-final for six extremely talented and intimidating British chefs as part of the Chef's Table round. The starter was complex: pan-seared scallops on brawn with Granny Smith apples, pickled turnips, a layer of ponzu and dashi jelly, bonito mayonnaise, wild puffed rice, shiso leaf and fresh nori.

At the time, I had no idea what bonito or dashi were, nor had I ever cooked with fresh nori. And, as I am sure you can guess, when you're in a competition as big as *MasterChef* there is nothing more terrifying than working with ingredients you've never heard of before. Relief came when I was summoned to prepare dessert, and one of my fellow contestants was handed the challenge of constructing this delicate starter. However, hungry for a bit more culinary knowledge, I paid attention to its preparation.

Behind the scenes, when the camera wasn't pointing in my direction, I nagged Sat and his team until they explained the unfamiliar ingredients and techniques to me. Being the truly wonderful people that they are, they talked me through the entire dish, explaining each element in turn: dashi was a Japanese fish stock used to make miso broth, bonito a dried fermented Japanese skipjack tuna and nori the edible seaweed species of red algae genus Pyropia.

I quickly learnt that what all of these ingredients had in common was their heavy umami flavour. If, Sat explained, I was able to harness the umami qualities of my dishes in the kitchen, this could give me a culinary advantage over my competition. It was the perfect advice at just the right stage of the contest, and it worked. Building the umami flavour into my own dishes through a more considered selection of ingredients was (I think) part of what helped me to go on and win.

Since then, I have looked at Michelin-star menus in a completely different light; it is easier now for me to recognise where chefs have cultivated the umami flavour profile to elevate their dishes to new heights. When you see a chef adding miso paste to their chocolate and butterscotch pudding, that's them infusing umami into their dish, taking it to a whole new level. I must admit I have latched on to this 'miso-dessert' trend and add a touch of miso to virtually all my sweet creations. It seems to not taste savoury as such – rather, the inherent sweet flavours seem somehow more well-rounded and intensified. If you try my Miso, Date and Dark Chocolate Cookies (see p. 95), you will come to a new understanding of what the term moreish means. They never seem to last very long. I also don't shy away from using a range of different miso pastes in my savoury dishes, and there are literally hundreds of varieties to choose from. My current favourite umami hit is a Korean fermented soybean paste called doenjang, that isn't dissimilar to miso: it is gloriously assertive in its umami flavour profile and pairs splendidly well with mushrooms, as you will see on p. 91, where I use it in an irresistible noodle dish.

Through my recipes, you will see that umami is not to be found only on fine-dining menus; rather, it is an extremely powerful tool that can be used to enhance the food that we make on a daily basis. I'm not talking about covering your food with shop-bought MSG powder, but about harnessing the natural power of umami that exists in the condiments that surround us. Many instinctive home cooks will have been doing this for years without even realising it. Adding a squirt of ketchup to cottage pie mince, or a dash of brown sauce to cheese on toast or perhaps a splash of Worcestershire sauce to Caesar salad dressing – these are all ways that we add umami into our cuisine. And knowing that you can use these condiments to augment umami notes and produce more delicious food is indeed empowering.

I am also certain that most of you will have a kitchen cupboard where the umami-rich foods will have clustered together in gastronomic harmony. In my store cupboard, the following condiments have found themselves next to one another by nature, rather than by design: tomato ketchup, soy sauce, fish sauce, peanut butter, tahini, miso paste, stock cubes, gravy granules, Worcestershire sauce, brown sauce, anchovies in olive oil, dried porcini mushrooms and, of course,

Marmite. These umami-rich foods somehow found one another in my cupboard and will always be together as long as they live under my roof. I get through these condiments faster than most, particularly my fish sauce which I use liberally to make a particularly addictive, lip-smacking Thai dipping sauce called Nam Pla Prik (see p. 90).

Umami is deeply entrenched in the flavour profiles that human beings crave – our very first food, breast milk, even contains free amino acids full of umami properties. An adequate supply of amino acids is central to an infant's growth, to the structural integrity of their body's cells and to the stable function of a wide range of essential enzymes and hormones. And the most abundant of the free amino acids present in breast milk, at around 0.02 per cent concentration, is glutamate. A baby weighing 5kg who consumes, on average, say, 800ml of milk a day, will take in around 0.16g of glutamate (about the same amount that can be found in approximately 65g of tomatoes).

Sweet tastes are vital for energy acquisition, while salty tastes are necessary for providing us with a range of important minerals. Bitter and sour tastes exist to detect poison and food spoilage. So, what beneficial role does umami play, from an evolutionary standpoint? If we are exposed to the chemical compounds that make up the taste of umami from birth onwards, then it could be reasonably assumed that they serve some essential bodily purpose. Some scientists think that umami allows the detection of sources of amino acids, thereby ensuring that we never become deficient in protein; or at least it's a way that our bodies can identify which foods are likely to be good sources of protein when we are running low.

Umami through the ages

Looking back through history, it is clear that umami-based condiments are not a new invention. Garum, the name given to a fermented fish sauce used in ancient Greece, Rome, Carthage and later Byzantium, was common in those cuisines for hundreds of years. Another similar fish sauce, called liquamen, was hugely popular in the western Mediterranean and in the Roman era, and appears in a variety of different recipes in the Roman cookbook, *Apicius*. One ancient recipe for lamb stew from this historical cookbook sounds quite nice. It calls for lamb chunks cooked with onions, coriander, pepper, lovage, cumin, wine, oil and liquamen, and it is thickened with flour before serving. I definitely wouldn't mind being served that in a restaurant.

Similarly, murri or almori was a condiment made of fermented barley and used in medieval Byzantine and Arab cuisine. It was used in small quantities, but in virtually every dish that we know of from that era. It took almost ninety days to make this paste, starting with the wrapping of barley in fig leaves, and ending with forty days of fermentation.

A bit later, in the late 19th century, the 'king of chefs and chef of kings', Auguste Escoffier, developed his then-famous veal-bone stock. He suggested that, apart from the four basic tastes, a distinct fifth taste was responsible for its mouth-watering properties. Unfortunately, most well-heeled Parisians didn't really care about why Escoffier's stock tasted so good; instead, they simply lined up around the block at the Ritz in Paris to sample its delights.

Today, in the small fishing village of Catara in Italy, you can buy *colatura di alici*, a fermented, amber-coloured anchovy sauce, which many think is a precursor for garum itself. And closer to home, we have Worcestershire sauce, a condiment whose original list of ingredients was rediscovered in 2009 and includes such umami gems as '8 gallons of soy' and '24lb of fish', and which still graces the plates of steak connoisseurs after all these years. It was created by chemists John Lea and William Perrins in the early 1800s, names which might ring a bell to Worcestershire sauce connoisseurs.

Worcestershire sauce has found its way into every strata of dining, from drizzling on top of cheese on toast in student halls to the kitchens of the finest chefs in the world. Marco Pierre White, one of the UK's

Worcestershire sauce – a happy accident

Legend has it that upon his return to Worcestershire, Lord Sandys, an aristocrat who had been the former Governer of Bengal, visited Mr Lea and Mr Perrins at their shop in Worcester. His request? To satisfy his appetite for a strongly flavoured sauce or chutney that he had tasted in India. Sadly, or so the story goes, the concoction the chemists came up with tasted vile at the time and was stored in the depths of the chemist's cellar out of the way. There, it slowly fermented and developed its umami properties until it was rediscovered some years later. Why exactly they chose to taste a years-old bottle of already vile-tasting sauce remains a mystery, but they found that it had transformed completely, mellowing to give the liquid that characteristic saline tang. Now, I wouldn't suggest that you should *always* taste barrels of decades-old liquids that you come across, but I guess what I am saying is that I'm thankful that these brave souls did.

best-known and most outspoken chefs, claimed that Worcestershire sauce allowed him to make 'the most delicious sauce in the world to serve with beef'. Nowadays, most people will have a bottle somewhere in their kitchen cupboard, even if it has been sitting there since they bought it to put on cheese on toast in 1998.

However, the biggest breakthrough in umami research came in 1908, when Dr Kikunae Ikeda, Professor of Chemistry at Tokyo Imperial University, watched his wife prepare dashi broth, a Japanese soup made from a dried seaweed called kombu. He observed that the kombu made meatless foods taste meaty somehow, and his curiosity led him to run the seaweed through various evaporation experiments.

After almost a year of research, he was able to isolate a crystal from the dashi, which he then tasted. It was the precise savoury, meaty flavour profile he was able to detect in his wife's broth. The crystal's molecular structure was identified (C5H9NO4, for anyone interested), and Dr Ikeda poetically and lovingly named it umami, a riff on the Japanese word *umai*, meaning delicious. He also, sensibly, patented the compound.

Like many scientific discoveries, Kikunae Ikeda's discovery went largely unnoticed at the time, failing to gain popular mass appeal in either the culinary or the scientific worlds. It clearly arrived before the world was ready for it (a bit like driverless cars, or Phil Collins), when most people were still unable to look at any seasonings beyond salt and pepper. Believe it or not, Ikeda's original paper, which was published in Japanese, was only translated into English in 2002, a full sixty-six years after his death, and only in the last two decades have advances in molecular biology allowed scientists to isolate the taste-bud targets responsible for the detection of C5H9NO4.

However, Professor Ikeda certainly had the last laugh. An entrepreneur by nature who quickly saw the potential of his compound, he capitalised on his discovery by mass-producing monosodium glutamate, or MSG, using fermented vegetable proteins. This was marketed as Aji-no-moto, the 'gourmet' powder (and Sharif's secret weapon) that has since been used by the food industry to build flavour in all manner of commercially produced foods. Ikeda passed away a rich man in 1936 and remains one of the greatest inventors in the eyes of most of Japanese people.

Commercial MSG: hero or villain?

MSG has been famous and infamous in equal measure over the years. It gained notoriety in 1968 when Dr Robert Ho Man Kwok, a senior research investigator at the National Biomedical Research Centre in the United States, wrote a short letter to the *New England Journal of Medicine* pondering the possible causes of some symptoms (numbness at the back of the neck, weakness and heart palpitations) he had after, he claimed, eating American-Chinese food in restaurants. In his own words, Dr Kwok noted that he had 'experienced a strange syndrome whenever I have eaten out in a Chinese restaurant, especially one that served northern Chinese food. The syndrome, which usually begins fifteen to twenty minutes after I have eaten the first dish, lasts for about two hours, without hangover effect.'

At the time, the use of commercial MSG in Chinese restaurants as an additive was reasonably widespread, and the theory that it might cause these symptoms caught on quickly. It led to the creation of what people called 'Chinese Restaurant Syndrome' (a term which nowadays can clearly be seen as inherently pejorative) and described a constellation of symptoms including headache, nausea, unusual numbness, and tingling after eating food in Chinese restaurants where commercial MSG was added.

Over the years, experiments to confirm the harmful effects of MSG in humans have mostly been carried out only on animals and have been inconclusive. One study, which found that the injection of huge quantities of MSG directly into the bloodstream of lab mice, resulted in some mice experiencing rapid brain deterioration, obesity and, in some, cases infertility; it caused a stir in many countries, but none of these effects has ever been seen in human test subjects.

So, just to be clear, studies looking at the existence of MSG's harmful effects on human health have not yet found any consistently reproducible negative effects on the human body. An average American consumes about 500mg per day of glutamate. That figure is higher in Asian cultures, but is nowhere near the 2,400mg *per kilogram* megadose that was used in the notorious mice study mentioned above. To reach that level, an average human (weighing 70kg) would have to consume nearly 168g of monosodium glutamate per day, which is a full *336 times* the average daily American intake of 500mg.

As with most things in life, context is paramount. A Chinese takeaway is overwhelmingly likely to have no negative effects, but you might see some if you sprinkle 168g of pure MSG powder on to your lunch, or start injecting it directly into your bloodstream (please don't do this).

Over the years, the hysteria around MSG has gone into overdrive. Some pregnant women avoid it on the basis that it may have negative neural effects on their newborns, while other people have blamed it, with little or no evidence, for a range of ailments such as weight gain, fertility issues or even the onset of Alzheimer's disease.

The issue, in my experience, is that MSG is often found in foods that are otherwise nutrient poor in order to make them more appealing. One of the things that makes MSG so valuable to cooks is that it modulates our perception of what could otherwise be quite dull ingredients. Classic examples are items like crisps, instant noodles and seasoned nuts. And, although MSG as a compound is man-made and does not exist in nature in its crystalline form, naturally occurring glutamate (and other amino acids which impart the umami flavour) are abundant in a range of highly palatable, nutritious foods such as asparagus, ripe tomatoes, cured meats, peas, walnuts, broccoli, Parmesan and Roquefort cheese, oysters and beansprouts to name a handful (see p. 85 for a table of umami-rich foods and their glutamate content). Even the process of slow-cooking a lamb shoulder will help release the naturally occurring, glutamate-rich, umami-tasting compounds from the meat.

I adore using these foods in my cooking and have designed some recipes through which you can delight in nature's umami-rich gifts. My Asparagus, Green Pea and Tahini Caesar (see p. 93) with its snow-like cascades of umami-rich Parmesan cheese tumbling over crisp lettuce leaves is a refreshing new twist on a classic. Umami-loaded tomatoes, bursting with sun-ripened sweet juices are found in a fabulous Turkish recipe for a spicy tomato salad which rather unusually uses both fresh tomatoes and tomato purée for added depth (see p. 94).

Through the recipes that follow it is clear that I am a believer in using the naturally occurring umami properties of food to create deliciousness in the kitchen. I also firmly believe there is more than enough evidence to consider umami one of our five basic tastes. I don't use commercial MSG powder in my own cookery, but that is because I prefer the inclusion of ingredients that naturally contain glutamate, and not necessarily because I think MSG powder is harmful to humans when used in sensible amounts. I certainly don't actively avoid all commercially produced foods that contain MSG, and I definitely will, on occasion, shovel a Chinese takeaway directly into my face.

As time goes on, I hope that the picture surrounding Dr Kikunae Ikeda's gourmet powder will become slowly, but surely, clearer. In the meantime, it's all about moderation, harmony and, of course, cultivating that sense of equilibrium with MSG. No doubt some of the most memorable culinary experiences in your life will have been realised through the clever use of umami-rich ingredients and condiments. Through this chapter, I hope to have empowered you not just to recognise the presence of umami in food that is presented to you, but also to have encouraged you to cook with the umami flavour in mind.

By considering the umami flavour in addition to salty, sweet, sour and bitter tastes, you will be able to harness maximal deliciousness in your food and push your gastronomy to new, unparalleled heights. In order to do this, I would encourage you to spend time familiarising yourself and experimenting with ingredients and condiments that are rich in naturally occurring umami; use them, as I have, to your advantage in your cooking. Remember, happiness in your kitchen is but a slather of Marmite, a sprinkle of Parmesan cheese, a splash of fish sauce and a tin of anchovies away!

Before you go on to browse through the umami-rich recipes that follow, I would like to end this chapter with a discussion of maybe the most important culinary creation ever: the classic burger. It contains the holy trinity of flavour combinations, namely perfectly cooked beef, melted cheese and slices of ripe tomato, and is something that I make often for myself and my family. It is a dish that brings me great joy to both create and eat. And I'm not alone, by any means. The answer to why this dish enjoys such global popularity lies, I think, in the fact that each of the three core ingredients are heavy in umami, and they combine within each mouthful to enhance each other's flavours. If you need to start your search for happiness somewhere, there are not many better places to begin than with the humble burger.

What's more, scientists have looked at the amount of dopamine (the feelgood chemical) that is released in the brain when we undergo various activities that bring pleasure. Dopamine creates a feeling of euphoria and satisfaction, and it is with delight that I can inform you that the research suggests that approximately two cheeseburgers can trigger the same amount of dopamine release as one orgasm. So, does one orgasm equals two cheeseburgers? Hmmm ...

Food	Free Glutamate Content (mg/100g)
Marmite	1960
Soy sauce	400–1700
Parmesan cheese	1680
Vegemite	1430
Roquefort cheese	1280
Dried shiitake mushrooms	1060
Oyster sauce	900
Miso	200–700
Green tea	220–670
Anchovies	630
Dry-cured ham	340
Tomatoes	140–250
Clams	210
Peas	200
Cheddar cheese	180
Oysters	40–150
Scallops	160
Shrimp	40
Corn	110
Potatoes	30–100

© Umami Information Centre

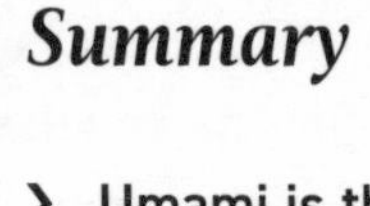

Summary

- Umami is the flavour reproduced when the taste buds detect glutamate. It has been used for centuries to boost the taste of many of our most adored dishes.

- Umami-laden foods can be extremely appetising, so spend some time checking out the glutamate content of your favourite snack foods. You might be surprised.

- Most people cook with umami without realising it. Think about some of the contents of your cupboards; it's likely that umami-rich condiments and seasonings line your shelves. These can be used creatively to enhance the flavour of your meals. Harnessing the power of umami in your kitchen will elevate your food.

- Consider the addition of umami flavours in sweet as well as savoury foods – a touch of miso goes really well with butterscotch, for example.

- There is a difference between commercially produced and naturally occurring MSG. While there is no real evidence to show that commercial MSG is harmful to humans, try, where possible, to stick to natural sources of umami, rather than using MSG powder.

The Classic Umami Quarter Pounder

Serves 4

One of the most enduring icons of American cuisine and culture, cooking the humble beef burger takes some practice. The addition of Parmesan, anchovy paste and Worcestershire sauce heightens existing umami notes in the beef, making it the ultimate irresistible comfort food. You can just as well experiment with miso paste, porcini mushroom powder or XO sauce in your burger patty should you wish.

Ingredients

500g minced chuck steak (75% lean, 25% fat, at least)
4 tablespoons grated Parmesan cheese
1 tablespoon Lea & Perrins Worcestershire sauce
1 heaped teaspoon anchovy paste
4 soft burger buns
4 teaspoons mayonnaise
4 teaspoons tomato ketchup
4 slices of strong Cheddar cheese
Salt and pepper
Vegetable oil, for frying

Optional extras:
Dill pickles
4 iceberg lettuce leaves, finely sliced
Sliced ripe beef tomatoes
Finely sliced onion rings macerated in lemon juice for 15 minutes

Method

1. Ensure that the beef is at room temperature. Combine the beef mince with the Parmesan, Worcestershire sauce and anchovy paste in a bowl and mix well. Divide the mince into four round burger patties, approximately 2.5cm thick and season liberally with salt and coarsely ground black pepper.

2. Toast your buns on a non-stick pan or under the grill till light golden. Spread mayonnaise on the base of the bun and ketchup inside the lid.

3. Now cook your burgers. Heat a non-stick pan over a medium high heat, place a burger patty into the preheated pan with a splash of vegetable oil. When the patty achieves a golden brown crust on the first side (approximately 2.5 minutes), flip it over and cook the other side for another 2.5 minutes until it forms a crust. I tend to cook just one or two patties at a time so that I can give them my full attention
4. Flip the burger once more and cover with a slice of cheese. Place a lid over the pan to melt the cheese, this will take around 30 seconds to 1 minute only.
5. Remove the lid from the pan and carefully place the patty on to the pre-sauced burger bun. At this stage you can add the optional extras – pickles, lettuce, tomato slices, onions, etc. – to your taste. Place the lid of the bun on the burger and serve immediately, ideally with crispy French fries as an accompaniment. They say heaven is a place on earth, right?

Cheddar, Marmite and Chilli Cornbread

Serves 4

An admission: I fed this cornbread to friends who *abhor* Marmite without them knowing! They loved the quintessential soul-food dish and asked for the recipe, at which point I had the particular joy of revealing that the secret ingredient is a teaspoon of the heavenly yeast extract. Mwah ha ha ... gotcha!

Ingredients

2 eggs
1 teaspoon Marmite
50ml vegetable oil, plus extra for greasing
450ml buttermilk or natural yoghurt
70g self-raising flour

1 teaspoon bicarbonate of soda
½ teaspoon salt
275g fine cornmeal
125g mature Cheddar cheese, grated
2 green chillies, finely chopped

Method

1. Preheat the oven to 180°C fan. Grease a 23cm-wide circular non stick cake tin with a teaspoon of vegetable oil.
2. In a large bowl, whisk the eggs and Marmite until frothy, then whisk in the vegetable oil and buttermilk.
3. Sift the flour, bicarbonate of soda and salt together in a separate bowl and add the cornmeal. Stir with a wooden spoon to combine.
4. Now pour the dry ingredients into the wet ingredients and give the mixture a really good stir to form a batter. Fold in all but a handful of the cheese and all the chilli, using a wooden spoon. Pour the batter into the cake tin and top with the reserved cheese.
5. Transfer to the oven for 30–35 minutes. The outside of the cornbread will be a deep, dark golden brown colour. Check the bread has cooked through to the centre by inserting a toothpick or skewer: if it comes out clean, the cornbread is ready.
6. I like to serve this cut into wedges with, you guessed it, more Marmite and butter on the side. Scrumptious.

Note: for another flavour and texture dimension, you can add a handful of finely diced peppers and sweetcorn kernels to the batter before baking, if you wish. If you don't have a non stick tin be sure to line your tin with baking parchment from the outset.

Pomelo, Mackerel Salad with Nam Pla Prik

Serves 4

Nam Pla Prik is a traditional Thai condiment that uses fish sauce as its show-stopping umami ingredient. A good fish sauce tends to be light brown in colour. If it is dark, or has turned black, it is old and should be avoided. Although you can buy fish sauce in most supermarkets, the very best stuff comes with a sizeable price tag attached.

Your best bet is to head over to the nearest Asian store where you can buy 720ml bottles, like Squid brand, for a very reasonable price. And while it may seem that 720ml is a lot, believe me, once you make Nam Pla Prik, you will not be able to stop. You will find yourself dunking spring rolls and salad leaves in it. Even your favourite curries, fried eggs, guacamole and soupy broths will not be spared the Nam Pla Prik treatment.

Ingredients

100g dry rice noodles
300g pomelo segments
250g hot-smoked mackerel
3 spring onions, finely sliced

For the dressing:
4 garlic cloves
2 tablespoons palm sugar
4 tablespoons fish sauce
6 tablespoons lime juice
3 green bird's-eye chillies, finely chopped
15g coriander, finely chopped
Salt, to taste

Method

1. Start by preparing the rice noodles according to the manufacturer's instructions. This involves pouring boiling water over the noodles and allowing them to steep for a few minutes until they have softened, taking care to not soak them for too long, as they will become sticky. Strain the noodles and discard the water.

2. Scatter the noodles over a platter and top with pomelo segments, bite-sized flakes of the smoked mackerel and the spring onions.
3. To make the Nam Pla Prik, place the garlic in a mortar and pestle and pound to a paste. Add the palm sugar and crush the sugar crystals into the garlic with the pestle. Add the fish sauce, lime juice, chillies and coriander and give everything a good stir to combine.
4. Taste the dressing, as you can adjust sweetness, sourness and heat to your preference at this stage. Fish sauce is salty, but you can add a small amount of salt to taste if you wish. To complete the dish, spoon the dressing over the noodle/mackerel/pomelo mixture and toss everything together gently. Serve immediately.

Koreo-Chinese Black Bean, Mushroom and Spinach Udon

Serves 2–4

Korean doenjang (soybean) paste was quite the culinary revelation for me. It is a bit like miso paste, but more assertive, sharper, deeper and more complex in its flavour profile. The combination with oyster sauce is a strange Koreo-Chinese mash-up that my mind has created for this completely and utterly addictive noodle dish. Feel free to use soba noodles, or even egg noodles if you wish. I use the pre-cooked udon, which comes vacuum-packed, to save time and effort as I can be exquisitely lazy at times.

Ingredients

1 teaspoon doenjang soybean paste
1 heaped tablespoon oyster sauce
2 teaspoons sesame oil
1 red chilli, finely chopped or 1 teaspoon red chilli flakes
1 teaspoon honey
3 tablespoons vegetable oil
200g mushrooms (e.g. shiitake or sliced button mushrooms)
1 teaspoon grated ginger

½ teaspoon grated garlic
1 x 240g tin black beans, drained
400g pre-cooked udon noodles
100g fresh baby spinach leaves

Method

1. Start by combining the soybean paste, oyster sauce, sesame oil, red chilli and honey in a small bowl to make a stir-fry sauce. Set aside.

2. Heat a wok to a high heat and add the vegetable oil. When the oil is hot, but not quite smoking add the mushrooms and allow them to fry off and brown over 3–4 minutes (the idea is that the pan is hot enough that the mushrooms are sealed and don't stew and release their juices).

3. Add the ginger and garlic to the mushrooms, followed by the black beans. Cook the ginger and garlic in the wok with the beans and mushrooms for a further 2–3 minutes. This step is vital to mellow the flavour of the raw garlic and ginger. You will need to keep stirring the contents of the wok to prevent burning. If the wok looks and feels too dry, you can add a touch more oil to help things along.

4. Finally, drop the udon noodles into the wok, followed by the stir-fry sauce that you mixed earlier and cook everything through for a further 2–3 minutes. A few splashes of water may be needed to rehydrate and soften the noodles at this stage. To complete the dish, turn the heat off and add the baby spinach leaves. They will wilt in the residual heat of the pan and require no further cooking. Serve immediately and delight in the deep, dark umami richness of this dish.

Note: there is no need to add any extra salt, as both the doenjang paste and oyster sauce are deeply savoury.

Asparagus, Green Pea and Tahini Caesar with Sourdough and Olive Oil Croutons

Serves 2 generously

The traditionalists among you may feel that the original Caesar salad should not be tampered with under any circumstances. I beg to differ. While I am an ardent fan of the classic variety, the addition of tahini imparts a rather pleasing sesame enriched umami depth. Peas, asparagus, anchovies and Parmesan are all naturally rich in glutamate, making this substantial salad a bit of a umami bomb, if I may say so myself.

Ingredients

2 garlic cloves
25g anchovies
2 tablespoons tahini
1 heaped teaspoon Dijon mustard
2 tablespoons olive oil
Juice of 1 large lemon
Salt and ½ teaspoon black pepper or chilli flakes

To serve:
2 large slices of slightly stale sourdough bread
Olive oil
125g cos lettuce
200g frozen peas, defrosted
200g asparagus
2 tablespoons, or more, grated Parmesan cheese
1 tablespoon sesame seeds, toasted

Method

1. Grind the garlic in a mortar and pestle until it forms a paste. Add the anchovies, crushing them to a paste with the garlic using your pestle. Add the tahini, Dijon mustard, olive oil and lemon juice and give the whole mixture a good stir to combine. The Caesar dressing will be quite thick at this stage, so slowly add warm water (a tablespoon at a time) from your kettle to loosen everything to your desired consistency. Adjust seasoning to your taste – I often add more salt and a touch of chilli flakes or black

pepper at this stage, but you may feel that the dressing is sufficiently savoury.

2. Heat your griddle pan to a high heat. Drizzle the sourdough slices liberally with olive oil on both sides and place it on the griddle for approximately 3 minutes each side, until crisp and charred. Cut the toasted sourdough into bite-sized chunks and set aside. Griddle the asparagus for around 3–4 minutes (the idea is to char the surface without making it soggy and wilted).
3. Roughly chop your lettuce leaves into bite-sized chunks and scatter them over a large platter. Top the lettuce leaves with the peas and grilled asparagus, followed by the dressing. Toss everything together lightly so the dressing coats the leaves and veg fully. Sprinkle over the Parmesan cheese, sesame seeds and toasted sourdough croutons. Serve immediately with extra dressing on the side.

Antep Ezmesi (Spiced Turkish-style Tomato Salad)

Serves 6–8

Tomatoes are nature's glutamate rich-gift, making them highly umami. The glutamate is mostly in the central portion of the tomatoes, i.e. the juice and seeds, rather than the flesh. Just think about the tomatoey flavour that erupts in your mouth when you bite down on a sun-ripened cherry tomato and all the sweet tomato juice floods your mouth.

It therefore seems a great shame to me that many recipes call for the seeds and juice to be discarded. Renowned Michelin-star chef Heston Blumenthal was one of the first to question the practice of discarding tomato seeds, and quite rightly so. Additional tomato purée is not essential in this recipe, but most definitely adds a certain crimson magic, enhancing the natural tomatoey tones. Try to pick one that is not too acrid and concentrated for the best result.

Ingredients

750g diced tomatoes (skin and juice included)
350g cucumber, deseeded and diced
100g spring onions, finely sliced
3 Turkish green chillies, finely sliced
25g parsley, finely chopped
1 teaspoon pul biber chilli flakes or ½ teaspoon of red chilli flakes
1 teaspoon dried mint
2 generous tablespoons extra virgin olive oil
2 tablespoons pomegranate molasses
1 tablespoon tomato purée
1 tablespoon white vinegar
1 teaspoon white sugar
Salt, to taste

Method

1. Place all the ingredients in a large bowl. Give the salad a good mix, ideally with your hands and season generously with salt. Allow the salad to rest for 10 minutes or so before serving with flatbreads to mop up all the sweet juices.

Miso, Date and Dark Chocolate Cookies

Serves 4 (makes approximately 10–12 cookies)

I realise that some of you will consider the addition of miso to cookies borderline transgressive. I assure you it isn't. To be honest, you can't really taste the miso, as such – rather, the other sweeter flavours of dates and dark chocolate are somehow accentuated.

Shiro miso, often called white miso, and the even milder and sweeter saikyo miso are probably the pastes of choice when experimenting with desserts. Add a little to your chocolate sauce or to your favourite brownie recipe. The possibilities are endless if you conceptualise miso as a 'salt replacer' and flavour enhancer. A word of caution: although the darker styles of miso used to make soup can be kept refrigerated for months, sweet white miso can deteriorate fairly rapidly once opened.

Ingredients

120g soft light brown sugar
1 heaped teaspoon white miso
60g butter, softened
½ egg, beaten
100g self-raising flour
6 pitted Medjool dates, chopped as finely as you can
60g dark chocolate, chopped into small chunks (or just buy dark chocolate chips)

Method

1. Preheat the oven to 180°C fan.

2. Cream together the soft light brown sugar, miso and butter using a wooden spoon, until the butter looks slightly paler than when you started mixing. Beat in the egg: don't worry if it looks curdled at this stage, it will come together.

3. Sift the flour and add it to the egg mixture, together with the dates and chocolate chips. Get your hands into the mixture and combine everything to form a soft dough. Transfer to the fridge to chill.

4. After 20 minutes, break off walnut-sized pieces of the dough and roll into balls. Place these on a baking sheet, leaving space between them as they will spread during baking. Flatten the balls gently with your fingertips. This will help the cookies flatten somewhat more evenly as they cook.

5. Place in the oven and bake for 8–12 minutes, or until the cookies are dark golden on the edges and still look a little squidgy in the centre. As the cookies cool, they will become firmer in texture. After a few minutes of cooling on the baking tray, transfer them to a wire rack to cool completely, using a palette knife.

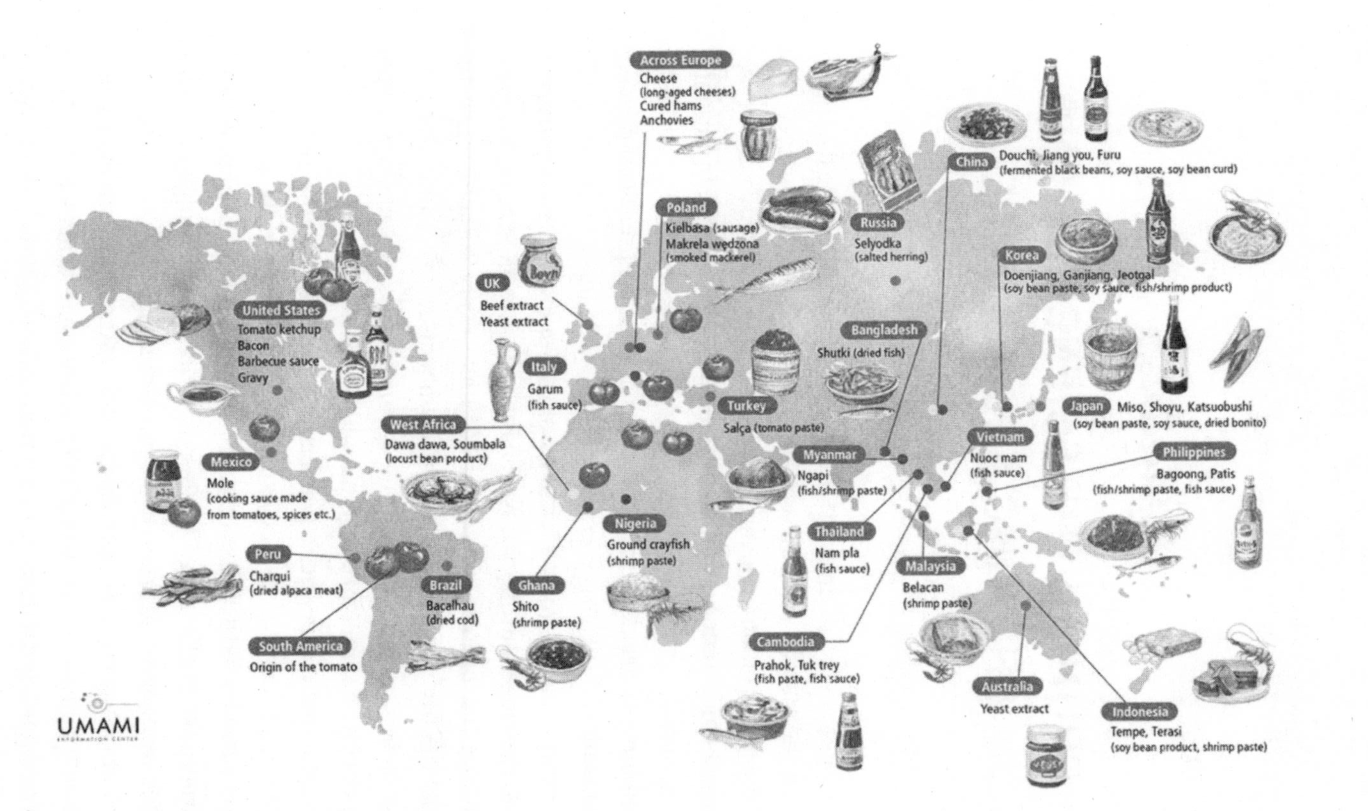

World Umami Map © Umami Information Centre

Chapter 4

CRUNCH: THE SCIENCE BEHIND CHEWING AND TEXTURE

Imagine a steak dinner, artfully cooked so that it's browned perfectly on the outside and pink and juicy on the inside. It's topped with crisp fried onions, and next to it is a pile of skillfully prepared triple-cooked chips with a a mound of al-dente charred green beans and a rich, smooth meaty jus brings everything together. Each mouthful is a mixture of heavenly tastes and varying textures that complement one another.

Now, imagine you chucked all of this in a blender and turned it into a homogenous, brown paste. The tastes are all exactly the same, nothing has been removed or added and every component of the meal has been cooked for the same amount of time and in the same way as before. But I assume you'd want your steak and chips served separately, not mulched together into a meaty smoothie ...

This small thought experiment demonstrates how important texture is to our overall food experience. It doesn't really matter how tasty a meal is (well, it does, but you know what I mean) or how many flavours are working together to create the most amazing taste if that other piece of the puzzle – texture – is missing.

Lack of texture is why meals with incredible ingredients and flavours can fall short, while fairly plain ingredients can be transformed into a great meal by a chef who understands the role of texture within the culinary experience. If you consider your own cooking in the last week, there may be countless examples where textural variation in dishes has amplified a meal to something great: a scattering of croutons on a silky, smooth soup or a handful of toasted peanuts tossed over your noodles are just a couple of examples of variations to be found literally everywhere.

When someone asks what your favourite dish is, if you are anything like me, your brain will go from one amazing dish to another, running in circles, unable to choose from the hundreds that you might want to pick. As a chef, I have been asked this question a number of times, and the answer is never straightforward, but if I had to choose my absolute favourite – the one that brings me the most joy – it would be the golguppa.

For those who may not have heard of this delicacy, the golguppa is an experience that everyone should partake in at least once in their life. It falls under the banner of 'chaat', the savoury street food served in India and Pakistan. Chaat carts, always colourful and rickety, are the melting pots that bring all demographics together, for there you will find street sweepers rubbing shoulders with CEOs and schoolchildren munching away on their after-school snacks alongside delivery drivers and couriers. And while you might not have a chaat stand on the corner of your road, your local Indian restaurant might have golguppas hidden on a menu under a different name – and even if they don't, they might be able to make some up for you to taste (if you ask nicely).

When I was eight years old, my family and I travelled to India for a holiday. Winding through the bazaars of inner-city Delhi, I dutifully followed my mother as she searched for the fabrics and handicrafts that made up her shopping list. It was midday, and the sun, already scorching hot, was beating down upon our heads with extra ferocity.

The great thing about the bazaars of Delhi is that you are never more than ten steps away from food. Roadside stalls line the streets in such a way that you have to go *through* the food to get to the actual stores.

Shops, kiosks and wheeled trolleys line up one beside the other, some selling fried savoury snacks like samosas, others filled with brightly coloured sweets or ice-cold drink bottles covered in condensation, all the scents mingling with each other in the hot air before being drawn into the nose.

A good rule of thumb is that the busier the stall, the better the food they are selling and, on the day in question, one stall in particular caught my mother's attention, with its long queue of customers, waiting to get hold of whatever was on offer. I say 'queue', but it was more like a cloud of customers, a haze of moving bodies with no discernible order, yet where each individual was completely aware of their place in the 'line'.

I watched the vendor as he served a customer ahead of us. A large pile of cylindrical, crisp, puffed dome-like shells lined a glass container, which he had placed on the corner of his cart. Without looking, he grabbed five shells in one hand, and placed them on a shallow plate made of recycled newspaper. He skilfully pierced each shell with a firm tap from his left thumb, while his right hand stuffed them with a mixture of spiced chickpeas and potatoes.

To his left was a large earthen pot, covered with a piece of netting (apparently, to ward off flies and dust, though how effective it was is up for debate). With a deftness and skill that come only from doing the same action a thousand times, he dunked each stuffed shell into the clay pot, one after the other. There were no gloves in use here, just two hands working expertly together to meet the demands of the eager line of customers. Each crisp shell was then submerged in a pale green-brown-coloured solution, before being topped with a bright green chutney and a mystery spice powder. Once the ensemble was complete, the vendor handed the plate to its new owner and turned his attention to the next person in line.

Each person in the queue in front of us stuffed shell after shell into their mouth, their eyes displaying their delight, even as their mouth remained busy. It is unusual to stand on a Delhi street and not hear anybody talking, but here the sounds of conversation had been replaced with the sounds of eating. These little balls of food seemed to be enough to make the crowd forget about the Indian summer sun, the traffic, the smog. For a moment, all that mattered were the morsels of food on people's plates and in their mouths.

'What is this, Mama?' I asked. It was nearly our turn and I couldn't wait to see what all the fuss was about.

'These are golguppas,' proclaimed my mother, her eyes fixed on the prize ahead.

Golguppas, as she explained to me, were a popular street snack in this part of the world, which she and her friends would often eat after college while waiting for the bus home. 'The most addictive food to exist on the face of this planet,' she told me. If my mother, usually such a harsh critic of any food served to her, was so complimentary, I couldn't wait to get my hands on some.

'But I don't think you or your sister can eat these today – you'll probably get ill. The water used isn't clean enough, and I really don't want you girls to fall sick on our vacation.'

Excuse me, what?

I looked at my dad, with all the pleading, puppy-faced anguish I could muster. He looked at my mother and decided that agreeing with her was less trouble than taking my side. They were both doctors, and feeding their children suspicious food from a Delhi street vendor was clearly not on their agenda. But still, the betrayal stung.

I went on the offensive. 'Why are you two going to eat them, then?' Within me, I could sense a rising anger and sense of disbelief at how unfair the situation was.

My mother mentioned something about having eaten these snacks through their entire childhood, and how their stomachs, unlike ours, were used to the risk of Indian street food, but I had stopped listening.

'That is really not fair,' I protested. All they had done was shop, dragging us behind them in the blazing heat. We deserved restitution. We deserved golguppas.

The vendor, a savvy entrepreneur who missed nothing, had been carefully watching me as I created a ruckus in the middle of the road. 'Don't worry, ma'am,' he said, before my mother had a chance to say a word. 'This is the cleanest golguppa stand in Delhi. No, in the whole world, actually.' He looked at her and continued. 'My customers are politicians, rich people, businessmen from the city and even babies. No one ever becomes ill. I use only mineral water for my dishes, see here.' He pointed to the drums of mineral water lined up neatly in the lower portion of his cart.

I looked at my mother, then the vendor and then back at my mother. Something was about to shift; we were on the cusp of a change. I find that at these moments it is best to remain silent, rather than say anything to jeopardise your slim chances of success.

'OK, you can try one and see if you like it,' she said rolling her eyes as she handed the vendor a few rupees. 'Children these days.'

With a smile extending from ear to ear, I took from the vendor's hands a plate of five golguppas. Slowly, delicately, I picked one up. It felt odd in my hand, like it didn't quite know which way up it was supposed to orient itself. I watched my mother pick one up and, shielding her blouse from any drips with the paper plate, she placed the whole thing into her mouth and crunched down. I copied her, opening my mouth as wide as it would stretch.

I bit down. The crisp shell collapsed between my teeth and the mint-infused fluid flooded my mouth. It was cold, deeply spiced and full of tart tamarind notes. My little mouth was no match for the sheer volume of flavour in each bite, and the residual juice escaped from the corners of my mouth, dribbling down on to my chin. Swallowing the extra liquid, I felt a further crunch as the crisp shell and mellow softness of chickpeas and potatoes slipped down my throat. The juxtaposition of crunchy, watery and soft textures was quite extraordinary.

Eating a golguppa is not a cultured affair, and there are no airs or graces involved at any point in the process. Businessmen and women, nannies, shopkeepers alike – it doesn't matter if you're the queen of England: they all have sauce dripping from their chins and down their arms, and it's infinitely more enjoyable when you don't think about it too much and just relish the whole messy experience.

It's important to note, in the interest of complete honesty, that over the following two days I developed terrible diarrhoea, which I suppose is a small price to pay for such delights. But let it also be known that my parents, with their apparently infection-resistant immune systems from a childhood of eating street food, also shared the same fate. Ha ha!

If you think a few days on the toilet would put me off golguppas, then you don't know me very well. Now, whenever I see them, I will go out of my way to grab as many as I can. To me, it is not just that they taste fabulous, with a complex interplay of salty, tart, spiced and slightly sweet flavours working together in harmony. It is that when I eat them, I become aware of the joyous process of chewing, and the pleasure that comes with sensing two, three or more oral textures at one time.

It is this simple joy, and the understanding of how we can harness the power of texture in our day to day culinary lives to derive maximal pleasure from food that we will be exploring in this chapter.

Gol Guppas

Serves 4

The ultimate taste and texture experience. Crisp fried shells are stuffed with soft mashed potatoes and chickpeas before being submerged in a cold, spiced tamarind water.

Warning: you must swallow the entire gol guppa whole, so as to avoid the juice exploding all over you! For ease, I use ready-made gol guppas which are available in South Asian food stores, or of course via the Internet.

Note: gol guppas can also be called 'pani-puri'.

Ingredients

50 ready-made gol-guppa shells

For the gol-guppa water:

4 heaped tablespoons ready-made tamarind chutney
25g coriander
15g mint leaves (stalks discarded)
2 green chillies
1 teaspoon chaat masala
1 teaspoon black salt
Juice of 1 large lemon
1 heaped teaspoon sugar
800ml ice-cold water

For the gol-guppa stuffing:

2 medium-sized boiled potatoes (approximately 450g)
240g boiled chickpeas (this is the amount in a standard tin)
1 teaspoon chaat masala powder
1 tablespoon finely chopped coriander

Method

1. Start by preparing the gol-guppa stuffing: dice the potatoes into small 1cm cubes and mix with the chickpeas, chaat masala powder and chopped coriander. Mash everything lightly with a fork to combine and just break down the chickpeas slightly. Set aside.

2. To prepare the gol-guppa water, place all the ingredients in a blender and blitz well till everything is really well combined.

Pass the liquid through a sieve into a pouring jug and refrigerate until you are ready to use.

3. To serve, pierce the gol-guppa shells with your thumb and stuff with a teaspoon of the potato and chickpea mixture. Then, fill the shells three-quarters of the way up with the gol-guppa water and consume immediately.

The art of mastication

As I write this chapter, and we embark on an exploration of chewing (or mastication, as it is more technically known), I find myself sitting in a coffee shop opposite a gentleman who is eating a slice of buttered toast. I watch him in a way that is not creepy at all and very clearly for research purposes. Without much conscious thought, his jaws move rhythmically as he chews his buttery toast, and when he finally swallows, I observe the gentle upwards motion of his Adam's apple as the toast moves past his throat and down his gullet, ready to embark on its exciting journey through the bowel.

The process of chewing food is far more complex than it first might appear. It's so normal, so ordinary, so unconscious that no one would bother giving it a second glance. We all do it hundreds of times a day, and, because it is so common, we forget that chewing is actually an exquisite neuromuscular dance of almost mind-boggling intricacy.

When you pop that piece of toast in your mouth, there are three key objectives that chewing needs to achieve. First, it has to break the large bite of toast into smaller pieces. Then, it needs to grind the morsel of food down to increase its overall surface area. Our teeth and our jaw muscles work together to carry out these two objectives.

Next, the chewing process needs to combine the toast with a healthy amount of saliva. This not only starts the digestion process of carbohydrates and fats, but also moistens the toast and binds it into a slippery food bolus (which is the fancy word for a blob of food that can be easily swallowed). The tongue, facial muscles and cheeks help with the formation of the food bolus, which is then swallowed. Swallowing is given the scientific name of 'deglutition'. So, first you masticate your toast into a bolus, then you deglute it. Simple, right?

Well, actually no. Chewing is, in fact, one of our bodies' most efficient and impressive mechanical processes. The whole system of grinding food between teeth by moving the jaw up and down could have gone quite wrong in evolutionary terms. Have you ever thought about why we don't constantly bite down on our tongues with every morsel of food? Or why we don't just grind our teeth into one another every time we chew something firm? If we did, we would quite quickly grind all our teeth down to nothing.

The key lies in the fact that the jaw has the ability to accelerate, and to vary its acceleration depending on how firm it senses the food in the mouth to be. But to avoid a painful dental grind, this acceleration is always followed by a controlled deceleration at the exact instant before your teeth knock into one another.

The process of breaking down morsels of food by chewing is what allows us to establish its texture and, alongside the ability of the oral mucosa (the skin inside the mouth) to pick up tactile stimulus through a perceptual process called 'somesthesia', our mouths can create a very clear picture of the physical characteristics of the food that we are chewing.

Depending on those physical characteristics, food will need less or more chewing. As a quick example, think about the difference between eating puréed carrot and a fresh carrot stick. You don't consciously say, 'Oh, here's some carrot purée. I had best alter my chewing style.' But the mouth knows what to do, based on the information it receives from the food's texture. In fact, the texture of food they were about to eat probably provided our ancestors with a very accurate clue as to its freshness, which, in the absence of a fridge or a best-before date, would have been very useful. Take the firm texture of fresh meat compared to the soft, slimy texture of a cut that has been left out too long, for example.

Getting to know your spit

Most Western cultures look at saliva with at least a little unease, and in some cases, complete disgust. But, in fact, spit has been quite popular in the past. The ancient Greeks, for instance, loved it, and believed that it cured a range of ailments; even today some Greek communities still spit three times to bring good luck for new babies. And practitioners of ancient traditional Chinese medicine believed that saliva was derived from the same component as blood, and so could provide similar diagnostic information about an individual's overall health.

The production of sufficient amounts of saliva is indispensable for good chewing. We produce around 0.5 to 1.5 litres of it a day from the salivary glands, and some estimate that we produce around 20,000 litres in a lifetime – that's approximately sixty-six bathtubs full of saliva. But how do we actually make all this spit?

Hidden behind the scenes, clusters of specialised cells (called salivary acini) that look like bunches of grapes are busy doing all the work necessary to deliver saliva into our mouths when we eat. These acinar cells secrete a fluid made of water, electrolytes, mucus and enzymes, all of which flow into collecting ducts. In these ducts, the composition of our saliva is altered, and after a while, they will coalesce and empty into larger ducts that finally empty into the mouth through the largest ducts. It is a system much like small tributaries converging to form a river, eventually flowing into the sea. Except the sea is your mouth and the water is your spit.

Saliva has many jobs, but one of its lesser-known roles is that it protects the teeth as you eat. It contains an array of antimicrobial components including lysozyme, an enzyme which kills many types of bacteria and prevents the dangerous overgrowth of oral microbial populations. Even though it contains millions of bacteria, research is shedding new light on the antimicrobial properties of saliva and its potential disinfectant properties. It is also a natural painkiller, and contains a compound called opiorphin, thought to be stronger than morphine. We have also, since the first studies were carried out in the late 19th century, known about the ability of saliva to break down carbohydrates and proteins. The enzymes it contains (amylase and protease, among others) also function similarly to those found in some bioactive washing up liquids – but please don't spit on your dishes if you run out of Fairy Liquid.

But what happens if it we don't have enough saliva? Consider what it might be like to have a constantly dry mouth? Called xerostomia and arising from various conditions, it is incredibly difficult to live with. I have met patients who suffer with it and report 'dysgeusia', or disordered taste, as one of their many symptoms. Not having enough saliva can literally change how you taste, or even remove that ability entirely.

Clearly, we take our saliva for granted. The fact that our mouths fill with extra saliva when we sniff a cake baking in the oven (see box opposite), or when we see cheese melting under the grill, is evidence of its importance as the first thing our food meets on its journey through our bodies. So, don't think of spit as a social taboo; rather, think of it as an underrated fluid which enables much of the gastronomic joy in our life. Be at peace with your spit.

Snap, Crackle and Crunch

Now that we have explored how we perceive different textures, and the vital role of saliva, I think it is time to talk about sandwiches – or, to be more precise, the Runny Egg and Crisp Sandwich.

This should be constructed according to a very narrow set of specifications. The crisps must be inside, not alongside the sandwich. The

Salivation – why your mouth waters

The nerves that control saliva production are part of a reflex system. They fire without us consciously needing to think about it when we are eating and the smell, taste and even movement of jaw muscles can activate this reflex. The part of the brain responsible for the 'salivary reflex' is called the medulla oblongata and it also controls a variety of functions including sneezing and vomiting. When the medulla receives food stimuli, it sends signals via chemicals (called neurotransmitters) causing the salivary glands to spring into action, manufacturing saliva.

Some scientists believe that the reason why the sight and smell of food makes our mouths water may be to do with 'classical conditioning' in our childhood. For example, imagine you once ate the most delicious waffle covered in molten chocolate and cream as a child. An association between that eating experience and salivation was subconsciously made. Therefore, in adulthood, the smell or sight of waffles can produce a salivary response, even though you are not actually consuming the waffle at that precise moment in time.

But salivation is complicated. The more we like the food, the more we are likely to salivate. Those people who consider themselves 'unrestrained eaters' tend to salivate more than 'restrained eaters' – for example, those who are on long-term diets. There is some evidence that the hungrier we are, the more likely we are to salivate, and even the act of vividly imagining a favourite food can make some people's mouths water profusely. Thus, the relationship between food and salivation is certainly a complex one.

egg filling to crisp ratio should be roughly equal. The cress should be mixed into the egg filling, never sprinkled on top. The bread should be thick sliced, and white rather than brown or seeded. If all of this is done correctly, when you bite into your creation you should first feel the soft bread between your teeth, which then gives way to the firmness of the salty crisps and then, finally, the creamy egg filling should take over, with little bursts of bucolic flavour coming from the cress. It's a delicate symphony, and one that is worth taking time over as you will see in the recipe on p. 118.

I cannot imagine eating an egg sandwich in any other way, and this is because of my love of the texture of crisps. I have had many heated discussions about the pros and cons of stuffing sandwiches with crisps. There are those in my life who disagree with my love for this delicacy, and sadly we must agree to disagree (although I know I am right here!).

The 'crisp' texture is the one that I value the most. I am always trying to think of ways to add crispy notes to my own recipes, to bring in the textural variation that makes food so interesting to eat. I recall watching the late culinary explorer Anthony Bourdain consume a crispy deep-fried Cambodian tarantula, and surprisingly, my reaction was not one of immediate revulsion. Instead, I remember thinking that insects would be far more edible and appealing to the wider culinary world if they were always served crunchy and deep-fried, with a side of sweet chilli sauce.

When I think of some of the classically 'crisp' foods I adore, one of the first things that comes to mind is pastry. I adore rough puff more than most people would consider normal. I simply do not feel at ease unless my freezer harbours a sheet or two of Jus-Rol, in case of a puff-pastry emergency. The fact is, that puff pastry is so very versatile. From simple straws to open tarts, pies, pasties and sausage rolls, the possibilities are seemingly endless. The paper thin, delicate layers of pastry are also used to make my dessert of dreams: the mille-feuille.

Mille-feuille translates literally to 'thousand sheets' in French and is classically made from three layers of puff pastry sandwiched together with crème patissière and topped with fondant icing or confectioners' sugar. I have vivid memories of trying my first ever mille-feuille as a ten-year-old in a quaint patisserie in Calais in the summer holidays. The contrast of wobbly custard and deep golden pastry shards was hard to forget. My parents packed a box full to take with us on our onward car journey to Paris; my siblings and I had near enough demolished the entire box in the back of the car before even setting foot on Parisian soil. So, I had to share my recipe for a rose-scented strawberry mille-feuille with you (see p. 119). I cheat by using sweetened whipped cream instead of crème patissière, but the result is a dessert that delivers 100 per cent on the flavour front, is visually impressive and remains desperately easy to put together. That moment where you first cut through it with a sharp knife knife and the cream and fruit bulge out as the pastry snaps, is what food dreams are made of.

But, I admit that puff is not the only pastry that excites me. Filo finds itself in strong second place in the pastry rankings. I particularly adore cutting diamond shapes through a tray of crisp baklawa and then drenching the crisp filo layers in a thick, sweet cardamom-laced syrup. Filo or 'phyllo' pastry takes to being filled with a combination of spinach and feta cheese incredibly well; as the Greeks do in a classic dish called the 'spanakopita'. I have shared my trusted recipe for a spinach and filo pie with you (with the addition of soft leeks, chickpeas and lemon) on p. 121. The trick is to make sure you drain the spinach well and brush each sheet of pastry liberally with melted butter to maximise crispiness.

The techniques that chefs use to make food crispy, are potentially endless – from the multitude of pakora and fritter batters to feather-light tempuras and even coatings for fried chicken (a recipe I could not resist sharing on p. 123). Being able to make your food crisp is nothing short of an art form. But if you don't wish to experiment much with the different techniques for creating crispness, at least familiarise yourself with 'pane'. This, according to the *Larousse Gastronomique*, is where 'food is coated in breadcrumbs before sautéing'. Once you know how to do it, you may even find yourself pane-ing every second item in your fridge. I have included a retro-fabulous recipe for crumb-coated goat's cheese on p. 124 which you can use as a blueprint for learning how to coat any food item with a crisp layer of breadcrumbs successfully.

Texture and sound

The term 'crisp' was derived from the Latin *crispus*, meaning curled. In the 14th century, the adjective was used to refer to wrinkled or rippled things, and by the 16th century, it had changed to mean items that were brittle, hard or firm. It was only in the 19th century that the term 'crunch' as a verb came to be used in the English language, probably as an onomatopoeic word mirroring the sound created when a crunchy item snaps.

The food industry is, of course, fully aware of our love for crunch. Look no further than the sheer number of options available in the crisp aisle to see just how much we love this texture, not just in the UK but across the entire globe. What is odd, though, is that surveys have consistently shown that the majority of us place more importance on the taste, smell and even temperature of food than on the texture of what we are eating. But at the same time, try feeding someone a bag of soggy crisps and see what their reaction is.

The availability of terms to describe texture vary hugely across languages and cultures. For example, where the Chinese and Japanese have dozens of words dealing with the crispy attributes of food, the Spanish don't, sometimes relying on the French term 'croquante' instead. In English, we have two main terms to describe this texture: 'crispy' and 'crunchy'. While most of us use these interchangeably, some food researchers argue that they are in many ways quite different.

While a 'crisp' food is dry, rigid and, when bitten with incisors (our sharp front teeth), fractures quickly and easily with a familiar, high-pitched sound, a crunchy food is dense and textured, which, when chewed with the molars (our wider, flatter back teeth), undergoes a series of fractures emitting relatively loud low-pitched sounds. So, a tub of Pringles is crispy, but a raw carrot baton is crunchy. I can't say that I'm convinced, but I will be sure to conduct more first-hand research and let you know.

Sonic chips

There are few science experiments that make me as happy as the 'Sonic Chip' experiment, or as it is otherwise known, 'The Role of Auditory Cues in Modulating the Perceived Crispness and Staleness of Potato Chips' experiment. This groundbreaking piece of research, conducted in 2005 by Professor Charles Spence and Professor Max Zampini on a tube of Pringles, revolutionised how we view the crunchiness of crisps. It also won the 2008 Ig Nobel Prize for the most humorous yet thought-provoking academic research of the year in the field of Nutrition.

Pringles, by virtue of being so uniform, paraboloid and stackable in their construction, are excellent subjects for the study of texture. Spence recruited twenty volunteers willing to bite into nearly 200 Pringles of varying degrees of freshness (in the name of science), and played them modified crunching sounds through headphones, some louder, some more muffled, as they ate. He found that he was able to change people's perception of the crunchiness and freshness of crisps by boosting high-frequency sounds.

Now, this research is most likely not going to change the way we see food. But what is important is that it talks about sound as the forgotten partner in our perception of flavour and texture. Manipulating sound can transform our experience of food, via a concept known as 'cross-modal sensory interaction'. What this means is that if you change the input in one sensory realm (i.e. you play the sound of someone biting down on a fresh, crisp Pringle), you can influence and often override the perception in another (the soggy pringle you are actually eating).

We know (from being alive) that the brain integrates information from all five human senses to produce a complete picture of what we are experiencing, and our experience of food is no exception. There is so much more than just the taste buds at play here.

There are even soundtracks that you can listen to online give your bag of stale crisps the boost they need. I tried it with a half-eaten packet of crisps that my son left out the night before, and I was surprised to find, not just that it worked, but how much it worked. The high-frequency track made those soggy crisps far more palatable and, now that I think about it, it is not a surprise that even crisp packets are designed using materials that sound crunchy. When you pick a packet up between your hands, it immediately starts creating that crinkly sound, which acts as a primer to make the experience of eating what's inside even crunchier.

How to eat a Magnum

Another classic example of the importance of 'crunch', this time in the realm of sweet rather than savoury food, comes from the Magnum ice cream. Apparently, food researchers working with Magnum's producer Unilever were asked to respond to customers' complaints that the chocolate coating the ice cream fractured too easily and either fell to the floor or stained their clothes. I would argue that there is not a single person who has eaten a Magnum who has not also watched with dismay as half of the chocolate covering detaches itself and falls to the floor.

Allegedly, the product development team for Magnum set about trying to resolve the issue, and came up with a chocolate coating that stuck to the central vanilla ice cream core more effectively. But, importantly, the distinctive chocolate snap that customers experienced when biting into the chocolate, was lost. As loyal customers bit into their ice creams, a far greater number of people now began to complain that the texture was different. The crunchy snap was more important to them than the collateral loss of some of their chocolate.

Textural words

We know that certain terminology in our own language can foster very specific food associations. Words like 'tender', 'crisp', 'succulent', 'juicy' – they all conjure images of particular foods and their associated experiences. But what if I described a particular food as 'gloopy'? Does that make it more, or less appetising?

The Japanese language is crammed full of onomatopoeic sounds, which mirror the textures of the food they describe. These sounds bring me great joy, and I am certain they enhance the taste and texture experiences of those enjoying them. In Japan, food can be *tsurutsuru*(smooth and slippery), *paripari*(crispy), *saku-saku* (a different kind of crispy), *neba-ne*ba (sticky) or *nicha-nicha*, *gunnyari*, *torori*, *doro-doro* and *beta-beta* (different types of stickiness, viscosity or elasticity). It would bring me a great deal of satisfaction if, in the future, some of these Japanese words, with their incredible sensorial quality, were introduced into Western descriptions of food.

As we now know, a huge part of our perception of texture is derived from sensations originating in the oral cavity; and the lips – the gateway to the body – are the most external part of that cavity. It is, for me, a nice coincidence that, alongside keeping our food in our mouths, they are also used to sound out all of these fabulously descriptive words. Just try not to talk with your mouth full.

What role, then, does texture play in our ability to understand food? How does it help a food lover in their quest for digestive health and happiness?

Well, food texture is important. You don't notice it so much while it's there, although maybe if it has been exceptionally deployed it might register on your radar. But when texture is not quite right, you are unlikely to enjoy the food in question, no matter how good the flavour.

I for one cannot stand a soggy-bottomed pastry or a stale biscuit that has lost its crunch. Don't even get me started on soggy roast chicken skin.

Most of us spend time obsessing about the flavour and smell of ingredients, the finest aromatic spices, the best-tasting melting, smooth chocolate, pungent cheese, tannic wine and so forth – but how much time do we really spend thinking about optimum food texture?

In some ways, the lack of appreciation of food texture and greater emphasis on flavour and taste of food is not entirely our fault. This is because textural awareness is largely subconscious; our sense of touch and hearing, as well as our saliva, among other factors, all play an essential role in allowing us to establish and enjoy food texture. But bringing textural awareness away from your subconscious and into your conscious thoughts when you are cooking will help you to cook better and make you realise that flavour on its own is just not enough.

You can begin to experiment with texture by adding crunch to meals that might previously have been made without much thought, or introducing different layers of texture, so that your mouth can fully appreciate the joy of the food that it is sensing. For example, a scattering of toasted seeds over your lentils or some crisp fried onions and croutons atop a soup will elevate these dishes to new heights.

A helpful exercise is to reflect upon different meals that you have cooked this week and consider points where texture could have been enhanced. Could your macaroni cheese have had a crispier baked crust? Maybe your mashed potato could have been creamier and smoother? Or that cake you baked more moist? Would last night's salad have benefited from some crispy elements?

Acquiring a heightened sensitivity to food texture in the kitchen will help every food lover in their quest to create tastier food and bring countless culinary rewards.

Summary

- The texture of food plays a huge part in how much we enjoy eating it. Mixing textures, for instance combining a soft texture and a crunchy one, is particularly appealing to many. Use this to your advantage when designing your dishes.

- Saliva has a number of functions, but it is especially useful in helping the brain establish the texture of the food we are eating.

- Our sense of hearing plays a huge role in establishing food texture. Manipulating what we hear while eating a particular food can, in turn, affect our experience of its texture.

- The language we use to describe food can influence how we perceive its final texture. When describing your food to others, think carefully about the adjectives you are using. A few onomatopoeic words scattered into your description can make food sound that much more delicious.

The Runny Egg and Crisp Sandwich

Serves 4

This is meant to be dirty! Sloppy egg mayo filling, tumbling out of soft bread and crunchy crisps stacked inside thick slices of soft white bread. I am a purist when it comes to this recipe. I have tried versions with miso in the mayonnaise, or truffle with the crisps. Alas, nothing is quite as satisfying as the simple original version.

Ingredients

5 large eggs
4 tablespoons mayonnaise
1 teaspoon Dijon mustard (desirable but not essential)
Handful of cress
8 slices of good-quality, thick white bread
50g butter
2 packets of Walkers ready-salted crisps or Kettle Chips (or 1 sharing bag)
Salt and black or white pepper, to taste

Method

1. Start by bringing some water to a rolling boil in your pan. Place the eggs in the boiling water for exactly 6 minutes. Remove the eggs from the boiling water and set aside to cool.

2. While the eggs are cooling, mix together the mayonnaise, Dijon mustard (if using), salt (be generous), pepper and cress in a bowl. Once cool, shell the eggs and cut them into quarters. Using a fork, combine the eggs with the mayonnaise; this will break them down a bit further, but the idea is to still keep them somewhat chunky. The eggs will be very soft-centred after 6 minutes of boiling, so the rich yellow yolks will marble the fatty mayonnaise. A runny egg filling is what you are after.

3. Place 4 slices of bread on your chopping board and top them with equal amounts of the egg mayonnaise filling, followed by

a healthy mound of crisps. Take the last 4 pieces of bread and butter them. Close the sandwich by placing the top slice of bread on the crisps, butter-side down. Enjoy the sandwich whole, or if you want to hear that satisfying crunch of cracking crisps against oozing wet egg mayo filling, cut the sandwich into half; triangular sandwiches are better than rectangular ones somehow. Some leaking egg filling is also desirable.

Strawberry and Rose Mille-feuille

Serves 4–6

Layers of crisp, flaking buttery puff are the show-stopping element of this deceptively easy, yet highly impressive dessert. The crunch of puff pastry against soft cream and fruit is where the textural magic is to be found. The addition of rose takes you on a culinary trip from an elegant French patisserie to an exotic Arabian night.

Ingredients

For the pastry:
1 sheet of puff pastry
1 tablespoon icing sugar

For the cream:
600ml double cream
1 heaped tablespoon icing sugar
¼ teaspoon ground cardamom
4 tablespoons rose jam

For the fruit:
400g sweet ripe strawberries, sliced 5mm thick
2 tablespoons rose jam
¼ teaspoon ground white pepper

To decorate (optional):
1 tablespoon dried rose petals or a few fresh rose petals
1 tablespoon finely chopped pistachios or other nuts of your choice

Method

1. Preheat the oven to 180°C fan.

2. Line a baking tray with greaseproof paper. Cut the pastry sheet into 3 equal-sized rectangles. Take 1 tablespoon of the icing sugar and dust the pastry rectangles on one side through a small sieve, using half the icing sugar in the sieve. Place the dusted surface of the pastry rectangles face down on the lined baking tray. Dust the top surface of the puff pastry with the remaining icing sugar in the sieve.

3. Cover the puff pastry rectangles with another sheet of greaseproof paper and then place another heavy baking tray on top to weigh down the dough as it bakes. Place in the oven for 20 minutes. Remove from the oven, and take off the top baking tray and greaseproof paper. If the dough is not a deep golden brown, return to the oven for another 5 minutes, or until the colour is just right. Allow the pastry to cool completely on a wire rack. Make sure you handle the rectangles carefully, so they don't break.

4. Pour the double cream into a deep bowl and whisk it along with the icing sugar and ground cardamom until it is firm. Ripple through the rose jam and place the bowl in the fridge.

5. Macerate the strawberries in rose jam and white pepper. The white pepper will lift the flavour of the strawberries without tasting spicy (a little tip I learnt from Nigella Lawson). Allow to sit for 10 minutes before using.

6. To assemble the mille-feuille place one pastry layer on a serving plate. Top with a third of the cream (spreading using a palette knife), followed by a third of the strawberries. Carefully place the middle layer of pastry on top and repeat the layering process with a third of the cream, followed by another layer of pastry. Place the final pastry layer on the top of the stack, followed by the

remaining cream and fruit. You can pipe the cream if you wish, but spreading it with a palette knife also works perfectly well and gives a rather natural and charming result. For a final flourish, scatter some dried or fresh rose petals and the pistachios over the assembled mille-feuille. Serve immediately.

Note: if you can't get hold of rose jam in your local supermarket, look in a Turkish or Middle Eastern food shop. You can just as well use rosewater instead of rose jam, but you will need to increase the amount of sugar you add to the cream and strawberries instead.

Feta, Spinach, Chickpea and Golden Sultana Filo Pie

Serves 4–6

There is something spectacular about shards of filo pastry cracking against the sharp edge of your knife as you cut into this pie. You can play around with fillings; I have added leeks and chickpeas here. It makes a great lunch on a weekend and a fantastic dish to take to a summer picnic.

Ingredients

2 finely sliced leeks
1 tablespoon olive oil
1 drained tin of chickpeas
1 x 380g tin spinach leaves (not purée)
200g feta cheese
30g sultanas
Juice of ½ lemon
Zest of 1 lemon
1 fat garlic clove, crushed to a paste
1 teaspoon oregano or thyme
1 teaspoon chilli flakes
75g toasted pine nuts
75g melted butter
7–10 sheets of filo pastry
1 tablespoon black or white sesame seeds
Drizzle of honey (optional)

You will need a 23cm round, non-stick cake tin.

Method

1. Preheat the oven to 180°C fan. Fry off the leeks in the olive oil until they have softened; this takes around 5 minutes on a medium flame. Place the leeks in a bowl and allow them to cool. Drain the tin of spinach in a sieve and discard the water.

2. Take the spinach leaves and give them a really good squeeze to release any more water. You will be surprised at how much comes out. Place the spinach in the bowl with the leeks, along with the chickpeas, feta, sultanas, lemon juice and zest, garlic, oregano, chilli flakes and pine nuts. Give everything a good mix, using a fork to combine uniformly.

3. Brush the cake tin with a little butter, then lay on a sheet of filo pastry, leaving some overhang at the edges, Brush with more butter, then lay another sheet over the first at a 45-degree angle. Repeat with the remaining sheets, overlapping as you go, until they are all used up.

4. Spread the filling over the base of the pie and then gather in the filo edges to make a crust. If there is a central portion of filling that is not covered, simply take an extra piece of pastry, brush liberally with butter and crinkle it up roughly to plug the hole. Scatter over the sesame seeds.

5. Bake in the oven for around 30 minutes, or until the pastry is a deep golden colour. Allow to cool slightly before carefully removing the pie from the cake tin using a slotted turner or palette knife. If you wish, you can drizzle the pie with honey before serving.

Note: I use tinned spinach as it is inexpensive, and I always have some lurking around in my pantry. You can use fresh spinach also, but will need to wilt it separately in a pan, then cool and wring out the moisture until you have about 240g cooked and drained spinach leaves to work with.

Sesame Fried Chicken with Gochujang Mayo

Serves 4

There is something deeply satisfying about watching crunchy golden chicken rising from bubbling vats of oil. If you are a born-and-bred Londoner, you will know what I am talking about. Now, I am not advocating having this every day – it is fatty and salty – but there are times where nothing but this crispy fried chicken will suffice to replenish the soul.

Ingredients

300ml buttermilk
6 fat garlic cloves, crushed to a paste/ grated on a microplane
1 thumb sized piece of ginger crushed to a paste/grated on a microplane
2 teaspoons hot paprika
2 teaspoons cayenne pepper
1kg boneless, skinless chicken thighs
150g plain flour
75g cornflour (or rice flour)
125g sesame seeds (black and/or white)
1 heaped teaspoon salt
Vegetable oil, for deep-frying

Method

1. Combine the buttermilk, garlic, ginger, paprika and cayenne pepper in a bowl to form the marinade for the chicken. Place the chicken in the marinade, cover with clingfilm and allow them to marinate for 24 hours in the fridge.

2. Mix the plain flour, cornflour, sesame seeds and salt together in a shallow bowl. Remove the chicken from the marinade and dredge them in the seasoned flour.

3. Heat the vegetable oil in a deep-fat fryer to 175°C. Lower two or three pieces of chicken into the oil. Use tongs to rotate the chicken every 2–3 minutes or so. It will take 5–7 minutes for the chicken thighs to cook through fully. If in doubt, you can check that the chicken is cooked by inserting a temperature probe into the thickest part of the chicken; it should read at least 73°C. Drain the chicken on a wire rack placed on a baking tray which acts to catch any hot fat that drips off the chicken, and prevents it from becoming greasy.

4. Serve while hot and crunchy, sprinkled with a little extra flaky sea salt if you wish and a Gochujang mayonnaise (made with 6 tablespoons mayonnaise, 3 tablespoons Gochujang chilli paste, a drizzle of honey, the juice of a lime and a pinch of salt).

Crispy Goat's Cheese with Lashings of Honey and Pul Biber Turkish Chilli Flakes

Serves 4–6

Now, I know many so-called 'foodies' raise their noses at dishes with goat's cheese, citing its astronomic popularity in the late nineties. But these oozing, crisp bites are a timeless thing of beauty. Just wait until the salty cheese melts, encased within the crisp breadcrumb shell. I guarantee you will feel weak-kneed after eating them.

Ingredients

480g goat's cheese (ideally 4 x 8cm long logs, weighing 120g each)

4 heaped tablespoons plain white flour

2 large eggs, beaten

120g panko breadcrumbs

1 heaped teaspoon pul biber chilli flakes, plus extra for garnishing

Vegetable oil, for frying

1 tablespoon runny honey

Method

1. Cut the goat's cheese into rounds about 2cm thick. I tend to choose a goat's cheese that is not too soft and has been refrigerated to make it easier to work with.

2. You will need three shallow trays: one with the flour, one with beaten eggs and the last with the panko crumbs. Add the pul biber chilli flakes to the eggs and stir gently with a fork to combine.

3. First dip the goat's cheese rounds into the flour and dust off the excess. Next, pass them through the egg, making sure they are well coated. Shake off any excess egg and then pass the cheese through the flour once again. This will create a sticky seal over the goat's cheese. Finally, pass the cheese through the breadcrumbs, tapping gently to ensure that they are well attached. Place on a clean plate and transfer to the fridge until you are ready to fry. This is a messy business, so expect sticky fingers – but it is also well worth the effort.

4. When you are ready to fry, pour the vegetable oil to approximately 1cm depth in a frying pan and place over a medium heat. The oil is ready when a few crumbs of panko fizz when thrown into the oil. Shallow fry in batches of 3 or 4, for approximately 2–3 minutes on each side, or until the crumbs are golden brown and crisp. Use a metal palette knife to manoeuvre the goat's cheese in the frying pan, as a fork may perforate the coating and cause the cheese to escape. Once the goat's cheese is ready, remove it with a palette knife on to a plate lined with kitchen paper to drain off the excess oil.

5. To serve, place the crispy goat's cheese bites on to a plate, drizzle over the honey and dust liberally with extra pul biber chilli flakes. Serve immediately and watch the goat's cheese bites vanish faster than you can fry off the next batch.

Chapter 5

SPICE UP YOUR LIFE

You may by now have realised that a huge part of my personal definition of digestive health and happiness comes from the experiences that are taking place in my mouth when I take a bite of food. I love knowing how my senses are working together, complementing and amplifying each other to create a sensory experience.

Being from the Indian subcontinent, if I wrote a book about how we experience food and didn't include a chapter on spice, I am reasonably certain that my South Asian credentials would be quickly revoked. But the reality is that it has been decades, centuries even, since spiced foods were found only in Asia or Central and South America. Indeed, many of my American and European friends have a higher tolerance for (and derive greater enjoyment from) spiced foods than I do. I think that this shows just how much spices have integrated themselves into Western culture.

Without a doubt, the world would not have been explored as early and as thoroughly as it was, were it not for spices and their value. As a child, I owned a beautifully illustrated book about the travels of Christopher Columbus who, intending to sail east to India, accidentally sailed west and ended up on the shores of the Caribbean and South America. He was said to be voyaging *ad loca aromatum*, to the places where the spices are and, despite his inability to read a compass, he succeeded, returning to Europe laden with jewels like chilli, allspice and mastic. His voyages, and those of others before and after him, have, through their accumulation of a world of knowledge relating to spices, influenced the taste profiles of dishes that we still eat today. For example, what would the paella be without saffron? Can you imagine an apple pie without cinnamon? Is biryani still biryani without cardamom?

It is no exaggeration to say that where spices went, money followed. The relative value of nearly all spices in the early and Middle Ages meant that the spice trade literally redrew the world map and was the engine that drove the global economy at the time. The world has changed so much since then, but our insatiable appetite for spices was the driving force behind the creation of trade routes that today span the entire globe. And the world is shrinking now; today, you can go on the Internet while drinking your morning coffee, order some of the rarest spices in the world and tomorrow morning, someone will knock on your door and hand them over to you with a smile and a wave. Two or three hundred years ago, those spices might have been reserved only for kings and queens, with a corresponding price tag. If you think of it in these terms, you'll never take the pepper sitting on your dining table for granted again.

To be clear, in this chapter I will first talk about spice in general (the dried seeds, bark or bark resin, stigma and roots of plants picked at various stages of ripeness that are found abundantly in nature), how they are classified and how they confer medicinal benefits. I will then focus on one particular spice that enjoys global popularity: the chilli pepper. I will not be covering different herbs which, in contrast to spices, are green and leafy portions of plants. So, for those of you for whom spices are an essential part of your palate (and also for those folk who don't like them but are curious), I hope that this chapter on how they affect the body is informative and helps you to develop your understanding of how you can utilise the health benefits of various spices to your advantage. At the same time, the recipes that I have included here are designed to deepen your love for spice, helping you to harness their immense power in your own kitchens.

Spices – from the shelf to the pan

Of all the dishes that I choose to cook, it is the heavily spiced ones that I hold closest to my heart. Cooking with spice is an art form, requiring not just an understanding of what spices taste like by themselves, but also how they will taste when you combine them, as well as how they interact with the foods you choose to pair them with. I'd argue that cooking confidently with spice is the closest that chefs get to that sense of 'flow' that musicians and maths prodigies talk about, when your body takes over to create masterpieces without you needing to consciously think about it.

Being able to access that feeling of flow when cooking with spices takes lots of practice, and a good deal of trial and error. As you practise, you'll learn that some spices need to be toasted to get their best flavours out, while others benefit from being fried, that some have to be rehydrated, while others turn bitter and discolour at high temperatures and so on. It definitely didn't hurt growing up in a Pakistani household where spices were ubiquitous in our dinners, but really, in order to successfully cook well-spiced dishes, all you need is the desire to experiment with the contents of a spice rack and learn from your mistakes and successes. It is futile to claim to have created recipes that represent the whole spectrum of spice, but I have included a few treasured ones in this chapter which celebrate different spiced notes.

I adore the way that spices can help me bring the world to my dining table each evening. One night I am in Kashmir sniffing a saffron-scented pilau, the next I find myself in Hungary, enjoying chicken paprikash. Or I can jet-set off to India and find myself at a Mumbai market stall in the sweltering heat where a street food classic called Aloo Bun Chaat, laced with fiery spices awaits (see p. 157).

And while spices might seem daunting at first, I guarantee that after a few attempts at cooking with them, you will have discovered at least two or three combinations and techniques that you can use to, well, spice up almost any dish. If you are still petrified of accessing your spice rack, there is no shame in using pre-made spice mixes. Ras el hanout, for example, is a wonderful Moroccan spice blend that can be used in tagines, soups rice dishes and vegetables. Baharat spice mix is used in the Arabian Gulf region as a dry rub for grilling meat or seasoning for beef, lamb, chicken, seafood and vegetables. I have written a recipe for baharat chicken with apricots and almonds on p. 160 to introduce you to this versatile mix.

Harissa paste is another favourite ingredient of mine. The recipes that have been developed in the last decade celebrating its use are seemingly endless; just type in 'harissa recipe' to Google and explore. I feel harissa pairs exceptionally well with aubergines and have included a recipe for you to enjoy it on p. 159. Za'atar is another beautiful Levantine blend of dried spices (and herbs), popularised by the likes of Ottolenghi. Sprinkle it on roasted tomatoes or over your cucumber and cream cheese sandwiches. Try Mexican Tajín chilli and lime blend sprinkled over fresh pineapple and mango or Indian chaat masala over apples, bananas and chickpeas. The options, crossing geography and culture, are virtually endless.

The classification of spices

With the variety of spices that exists in the world, classifying them into anything resembling a system is a Herculean task. But thankfully, someone decided to do just that, and that someone is food scientist and author Dr Stuart Farrimond. Taking inspiration from the scientific world, he devised a periodic table of spices, assigning them to one of twelve key flavour groups, depending on the flavour compound (the chemical molecules which give spices their distinctive taste) most

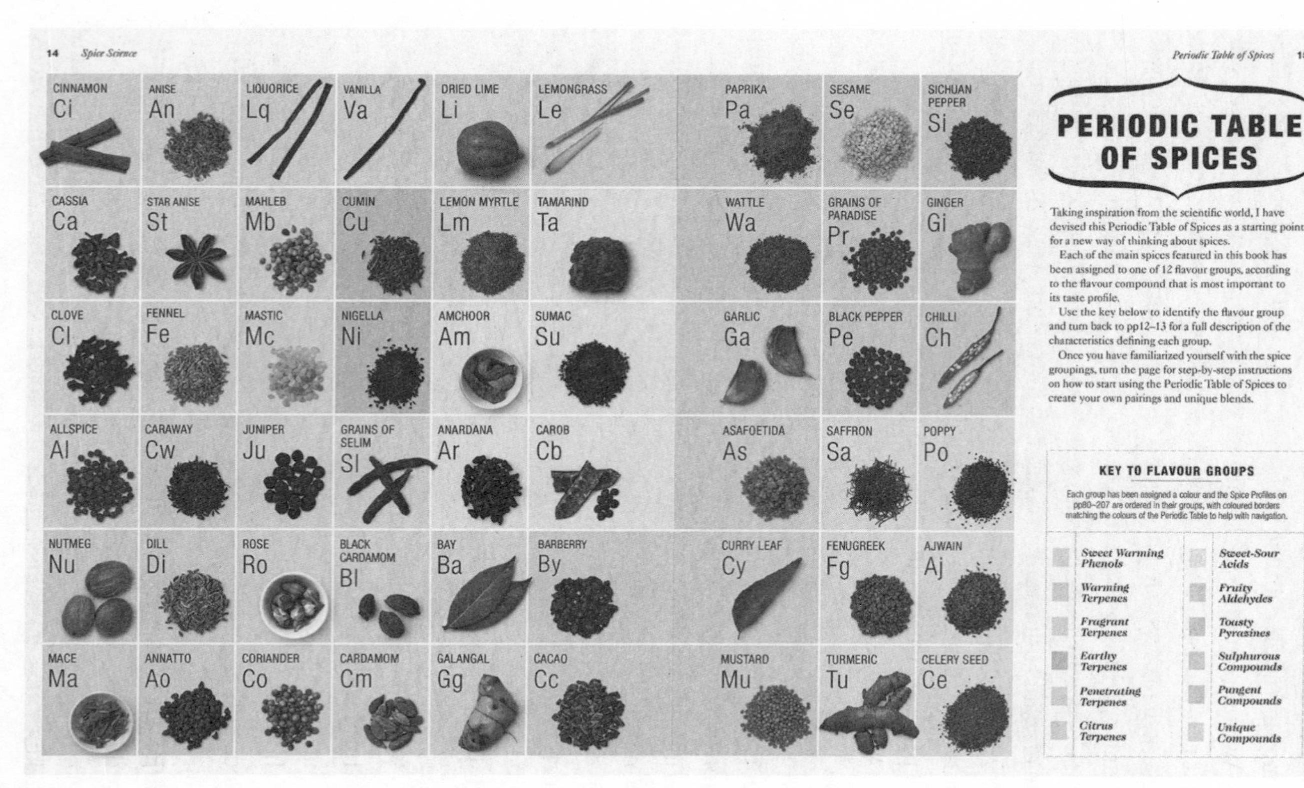

14 *Spice Science*

Periodic Table of Spices 15

CINNAMON Ci	ANISE An	LIQUORICE Lq	VANILLA Va	DRIED LIME Li	LEMONGRASS Le	PAPRIKA Pa	SESAME Se	SICHUAN PEPPER Si
CASSIA Ca	STAR ANISE St	MAHLEB Mb	CUMIN Cu	LEMON MYRTLE Lm	TAMARIND Ta	WATTLE Wa	GRAINS OF PARADISE Pr	GINGER Gi
CLOVE Cl	FENNEL Fe	MASTIC Mc	NIGELLA Ni	AMCHOOR Am	SUMAC Su	GARLIC Ga	BLACK PEPPER Pe	CHILLI Ch
ALLSPICE Al	CARAWAY Cw	JUNIPER Ju	GRAINS OF SELIM Sl	ANARDANA Ar	CAROB Cb	ASAFOETIDA As	SAFFRON Sa	POPPY Po
NUTMEG Nu	DILL Di	ROSE Ro	BLACK CARDAMOM Bl	BAY Ba	BARBERRY By	CURRY LEAF Cy	FENUGREEK Fg	AJWAIN Aj
MACE Ma	ANNATTO Ao	CORIANDER Co	CARDAMOM Cm	GALANGAL Gg	CACAO Cc	MUSTARD Mu	TURMERIC Tu	CELERY SEED Ce

PERIODIC TABLE OF SPICES

Taking inspiration from the scientific world, I have devised this Periodic Table of Spices as a starting point for a new way of thinking about spices.

Each of the main spices featured in this book has been assigned to one of 12 flavour groups, according to the flavour compound that is most important to its taste profile.

Use the key below to identify the flavour group and turn back to pp12–13 for a full description of the characteristics defining each group.

Once you have familiarized yourself with the spice groupings, turn the page for step-by-step instructions on how to start using the Periodic Table of Spices to create your own pairings and unique blends.

KEY TO FLAVOUR GROUPS

Each group has been assigned a colour and the Spice Profiles on pp80–207 are ordered in their groups, with coloured borders matching the colours of the Periodic Table to help with navigation.

Sweet Warming Phenols	*Sweet-Sour Acids*
Warming Terpenes	*Fruity Aldehydes*
Fragrant Terpenes	*Toasty Pyrazines*
Earthy Terpenes	*Sulphurous Compounds*
Penetrating Terpenes	*Pungent Compounds*
Citrus Terpenes	*Unique Compounds*

From *The Science of Spice* by Dr Stuart Farrimond © Dorling Kindersley Limited. Published by Dorling Kindersley Limited and reproduced with their permission.

central to each one. As a scientist and cook in equal measure, this brings me immeasurable joy.

When we swallow spices, their flavour compounds release aromas that rise from the back of the throat and climb up to the back of the nose, where, says Dr Farrimond, we experience them as if they were sitting right there on our tongues.

For those who are novices and wish to delve into the world of spice, the classification system (here) may prove helpful. By understanding how spices can be grouped together by flavour profile you will have the tools to start experimenting with them in the kitchen successfully – a deeply empowering position to find yourself in.

And for those who are already experimental with spice, the system can help you think out of the box. Can you replace a sweet-sour acid with a fruity aldehyde compound to spruce up a classic dish? Are there combinations that complement one another that previously never came to mind? When I am developing recipes, I enjoy browsing the periodic table of spices, pondering all those that can be paired together in magical alchemy.

Terpenes

The terpenes are the broadest and most common flavour compound. They can be divided into five groups:

1. **Warming terpenes** give food woody, bitter, peppery flavours, e.g. nutmeg and mace. These flavour compounds evaporate easily and, as a result, can be lost with prolonged cooking. When used well, however, they can transform meals. Think of the difference a touch of nutmeg makes to cauliflower cheese or apple strudel.

2. **Earthy terpenes** have a mellow, dusty, burnt flavour, and they give your dishes a woody spiciness – think cumin and nigella. The flavours are oil-soluble and linger on the palate. Try making earthy, cumin-laced kofta kebabs on homemade flatbread, studded with nigella seeds.

3. **Penetrating terpenes** hit the back of the nose and linger for some time, and can often have eucalyptus-like flavours, e.g. cardamom or galangal. They should be used in moderation due to their potentially overpowering taste. Cardamom and crème fraîche ice cream is perfect with some chopped toasted pistachios on a summer's day, but it is important to get the balance right. Too much cardamom will take over a dish; too little and it will taste of nothing.

4. **Fragrant terpenes** taste pleasantly fresh, pine-like or floral with woody undertones. They are fast-acting but short-lived, and don't tolerate long cooking times well, dispersing quickly in oil. Coriander seeds, juniper, mastic and rose find themselves grouped together in this category, and these flavours often predominate dishes in parts of the Middle East.

5. **Citrus terpenes** give food a tangy, refreshing, herbaceous aroma. Lemongrass is a classic example, as are the dried limes used in Persian stews to impart an earthy tartness in the background of the dish.

Sweet warming phenols

These are a group of strongly flavoured and potent spices with aniseed- and eucalyptus-like flavours and a slightly bitter edge, e.g. cloves, fennel and cinnamon. The flavours only reduce very slowly on cooking. Together with caramelised onions, these spices form the basis for one of my favourite comfort foods: lamb pilau.

Sweet-sour acids

These are water-soluble, fruit-based spices and are usually accompanied by sugars that amplify fruity tones and take the edge off strong, sour tastes. Spices that fall into this category include amchoor (mango powder), anardana (dried pomegranate seeds) or tamarind chutney, spices used extensively in street-food dishes like chaat across the

South Asian region. The classic chaat recipe calls for cubed potatoes and chickpeas to be dressed with amchoor and anardāna, sweetened yoghurt, lashings of tamarind sauce and coriander chutney and it is topped with fried crispy shards of gram flour.

Fruity aldehydes

These are found abundantly in fruiting plants like sumac and barberry. They have a fruity, malty, fresh flavour with sweaty nuances. This group is enjoying a renaissance and is now very fashionable in culinary circles. Sumac sprinkled over salad leaves is found everywhere these days, and barberry-studded basmati rice has begun featuring on many Middle Eastern-inspired tables.

Toasty pyrazines

Heated to temperatures greater than 130°C as part of their processing, these release most of their flavour when fried or toasted in a dry pan. They have nutty, caramel tones that are often accompanied by smoky, meaty, fresh bread-like flavours, like those found in sesame and paprika.

Sulphurous compounds

Highly pungent and dominated by oniony, meaty flavours with cabbage or horseradish tones, these can be slightly offensive in high concentrations, e.g. mustard or garlic, but when used in moderation, a spoonful of Dijon mixed into buttery new potatoes with a good crack of salt and pepper is pretty much as good as it gets.

Pungent compounds

These include all the varieties of hot chilli peppers. They are actually not flavours at all, but chemicals that hijack certain pain nerves that would normally send warning signals to the brain. These pungent compounds, like piperine in black pepper and capsaicin in chillies, give us that kick that we experience when eating spicy foods.

There are also a few **unique compounds** that do not quite fit into any of the other groupings, e.g. saffron or turmeric. They usually partner well with a variety of other spices, bringing distinctive aromas to a dish. I find that a little can go a long way when cooking with these spices, and I'm often bemused when a recipe calls for an entire teaspoonful of turmeric in a dish that serves four. Half a teaspoon is more than sufficient to impart colour and flavour to any dish for less than four people, without being overpowering.

When the spice cupboard meets the medicine drawer

We know the transformative effect that spices can have on our cooking, but what effect (if any) do they have on our health? Spices have also been used for centuries for their medicinal properties. The Ancient Egyptians used them to embalm their dead: cumin and anise were used to rinse the innards of the dead, and Tutankhamun's tomb was even found adorned with black nigella seeds and garlic. Alexander the Great was said to soak in baths tinted with saffron to heal battle wounds, while Greek physician Hippocrates made note of several hundred different medicines in which he credited spices for their curative effects.

Although the historical use of spices for medicinal purposes is widely acknowledged, unfortunately the efficacy of these compounds when used in our diets today has not been studied as extensively. As a doctor and a chef, I feel I might be well placed to say that given their potential antioxidant, anti-inflammatory, anti-carcinogenic and glucose- and cholesterol-lowering capabilities, it seems a shame that we do not use spices and diet in a way that complements our use of allopathy (i.e. regular medicine).

Many recent consumer studies have indicated that public interest in learning about spices and their health benefits is rising, and that, in certain situations, people would be more open to taking these than conventional medications, due to the perception that natural ingredients might have fewer side effects. In my work, I have found this to be true of many patients with a variety of ailments. So, here, we will focus on three spices with the most promising potential for health-related benefits: cinnamon, turmeric and ginger.

Cinnamon

I thought I would start the discussion about cinnamon with some nerdy trivia. There is evidence that nearly 3,000 years ago, Arab traders would tell stories of a fearsome 'Cinnamologus', or 'cinnamon bird', to deter potential competitors from searching for the source of the spice: they claimed that the beastly bird made its nest out of sticks of cinnamon, and in order to get their hands on the spice, traders would need to tempt the bird from its nest with chunks of meat. Movement of the Cinnamologus bird from its nest would make small twigs of cinnamon fall to the ground, then the quick-witted traders would perilously gather the precious spice sticks and bring them to trade in markets.

Now, this tale may sound silly, but rumour has it that for hundreds of years the ancient Greeks and Romans were scared away from the spice route by the tale of the fearsome Cinnamologus. Thankfully, cinnamon is available on practically every corner today and has formed part of our culinary repertoire almost globally.

I, for one, adore a good cinnamon bun. The swirls of soft white bread, layered with sweet-smelling cinnamon, covered in sticky, white icing. Cinnamon is a very emotive spice, marking festivity and warmth. The scent of cinnamon wafts through the air at Christmas as mulled wine, mince pies and Christmas pudding adorn the table. But even though it tastes fantastic, does cinnamon confer any health benefits? Could the humble cinnamon bun actually be good for you in any way?

There is evidence that extracts of cinnamon have antibacterial properties, attacking major respiratory and gastrointestinal pathogens. This has been verified in laboratory testing with animals – not in human trials, but results are encouraging. Cinnamon is also thought to have anti-inflammatory and antioxidant properties (a chemical substance that prevents or slows down the damage that oxygen does to cells) and small studies show that it may help mop up harmful free radicals (unstable molecules that can cause damage to the cells in your body). Cells damaged by free radicals can result in diseases like diabetes, hardening of the blood vessels and high blood pressure, so the health benefits of cinnamon, if verified, are not insignificant.

Cinnamon supplements added to standard diabetes medications, alongside other lifestyle therapies, have also shown some success in controlling blood-sugar levels in diabetics. (But because of the high fat and sugar content I would probably still refrain from having cinnamon in mince pie, Christmas pudding and apple pie format all the time if you can!) Liquid extracts of cinnamon can inhibit the activity of 'tau proteins', a key ingredient responsible for the development of Alzheimer's disease, but this hasn't been proven in humans, just in laboratory-based studies.

I am not trying to say that cinnamon (or any of the other spices in this chapter) should be used exclusively to treat human ailments, but rather that there is a huge amount of apparent potential that we should investigate further. It is hard to draw any definitive conclusions about the benefits of cinnamon at this stage, but on the basis of the limited research available, initial findings do seem favourable.

So, my view is that a bit of extra cinnamon in the form of a tea or a supplement (or, dare I say, a cinnamon bun every now and then) is overwhelmingly likely to do no harm, and may quite possibly come with some very welcome health benefits. Give it a try?

Turmeric

There is something so alluring about the bold, sun-tinged amarillo hue of turmeric (apart from when it stains your fingertips for a fortnight, that is). I don't think I am wrong when I say that of all the spices, popular media has latched on to the potential health benefits of turmeric most fervently.

There is strong evidence that turmeric exhibits powerful anti-inflammatory and antioxidant activity, and mops up free radicals. Turmeric supplements can reduce inflammatory activity by inhibiting the substances in our bodies that cause inflammation. For instance, they have shown some limited success in patients with inflammatory bowel disease and may lower the severity or frequency of inflammation episodes.

Studies, in the laboratory as well as on animals and humans, have shown that turmeric can have a number of benefits for those with issues related to cardiac function, vascular health and cholesterol. Pilot trials of turmeric supplements to treat irritable bowel syndrome (IBS) symptoms showed that two-thirds of people taking the supplements felt some improvement in symptoms after eight weeks of taking them. Another study looked at turmeric supplements in treating patients with osteoarthritis. Patients who took 2g of curcumin (a chemical in turmeric) a day for six weeks reported similar pain-relief effects to those who had been given 800mg a day of ibuprofen, which, if you've ever seen how quickly 800mg of ibuprofen can get rid of a bad headache, you will recognise as a hugely positive result.

While there is mounting evidence that turmeric has a number of beneficial properties, its use hasn't quite reached clinical practice yet. However, you won't do yourself any harm by getting ahead of the game. And you don't need to chug down one of those vile turmeric lattes to get your daily dose. Try some grated fresh turmeric in your eggs. Or, blend turmeric into yoghurt with some olive oil, salt and honey to make a gloriously sunny salad dressing or just add to hot water with some ginger and lemon to brew a turmeric tea. I love adding turmeric to my version of soothing chicken soup. It makes me feel infinitely better when I have a nasty cold and I have included the recipe here for you (see next page). A note of caution: please don't replace the medicines your doctor has prescribed with turmeric, but the overall message is that adding a little more of it to your diet is unlikely to do any harm, and will most likely do you some good.

The Anti-Inflammatory Chicken Soup

Serves 4

Many of you will already have a coveted recipe for chicken soup that has been passed down the generations in your family. It is indeed one of the most pleasing and painless meals to cook from scratch. Sadly, manufactured varieties will simply not suffice. Sorry, Mr Campbell.

The addition here of turmeric, famed for its anti-inflammatory properties, along with fibre-dense nuggets of pearl barley makes this recipe particularly special. It gives me great joy to share that whenever I have made this for friends and family members who are poorly, they have felt much better for eating it; even my husband with his man flu! Which is saying something.

Ingredients

1 medium onion, finely sliced
2 tablespoons extra virgin olive oil
1kg chicken, jointed and skin removed
1 teaspoon turmeric powder (or 1 teaspoon grated fresh turmeric)
150g dried pearl barley
½ teaspoon cumin seeds
1 teaspoon fennel seeds
1 bay leaf
6 garlic cloves
1 thumb-sized piece of ginger, chopped into big chunks
1.2 litres light chicken stock
1 heaped tablespoon cornflour (optional)
Juice of 1 lemon
½ teaspoon pul biber chilli flakes
Handful of finely chopped flat-leaf parsley
Salt, to taste

Method

1. Brown the onion in the olive oil in a large pot until it is light golden in colour; this takes around 6–8 minutes. Now add the pieces of chicken and allow them to seal and brown all over

(around 5 minutes). Add the turmeric, pearl barley, cumin and fennel seeds, bay leaf, garlic cloves and ginger to the chicken and onions. Top with chicken stock and bring to the boil. Simmer gently over a medium heat for around an hour or so. If the pan starts looking a bit dry, top up with small amounts of water from the kettle. After an hour turn off the heat and allow the contents of the pot to cool.

2. The next stage involves using your hands. Take out the cooled pieces of chicken and shred the meat off the bones. Be careful not to shred too finely at this stage as the chicken fibres will disintegrate further on cooking. Discard the bones and return the chicken meat to the pot. Discard the bay leaf and chunks of ginger. Top up the pot with around 500ml of warm water to bring it back to a soup/broth consistency and bring everything to the boil.

3. At this stage you have two options: you can leave the soup as it is, or if you prefer a thicker consistency, dissolve the cornflour in some water to form a cream-like solution and stir this into the soup. After a few minutes of boiling you will notice that the soup has thickened significantly.

4. Season with salt to taste (be aware that some chicken stocks contain salt already so you may not wish to add too much), then squirt in the lemon juice and sprinkle in the chilli flakes and a handful of parsley before serving.

Note: the stock is not essential – you can use water instead. The use of stock will, of course, intensify the flavours of the soup. You can substitute red chilli flakes for the pul biber flakes (which I love) or fresh red/green chillies to your desired level of heat.

Ginger

Ginger is one of those things that everyone's grandmother, at one point or another, has told them would cure their upset tummy. And actually, your gran might not have been wrong. Recent studies have suggested that ginger is a safe and effective way of reducing nausea and vomiting in pregnancy, as well as the nausea experienced during chemotherapy. And given that these are two of the most nausea-inducing things that a body can go through (apart from someone pronouncing espresso as *expresso*), that's pretty strong evidence in support of this root.

I originally learned about the beneficial anti-emetic (nausea-preventing) properties of ginger during my first pregnancy. I had terrible morning sickness, and the putrid scent of ... well, poo, on the gastroenterology ward where I was working would make me vomit each day without fail. A nurse, to whom I will be forever indebted, suggested that I try putting a small piece of ginger between my back teeth and gums to control symptoms. I was sceptical at first, but of course it worked, didn't it! No more running to the loo to be sick halfway through ward rounds.

More clinical trials are necessary to draw any definitive conclusions about the cardiovascular benefits of ginger, although early studies suggest it may prevent our blood platelets from clumping together, a cause of heart attacks.

I adore the peppery heat of ginger – fiery when raw but mellowing into a warm hug upon cooking. Ginger is also very versatile, combining with both sweet and savoury taste profiles with equal success. Crunchy ginger biscuits were a favourite of mine growing up (now I crumble them up into ice cream, along with extra pieces of candied ginger and ginger juice), but ginger is equally at home in a Thai green curry or an Indian savoury.

The story of capsaicin

Growing up, Friday night was my favourite night of the week. Bedtime was as late as we liked, and finally, after a tough working week, our whole family was able to sit and share some time together. My parents, both NHS doctors, were invariably exhausted, and cooking for two adults and three children was high on the list of their least desirable things to do. Takeaways were never an option because my mother, a traditionalist, felt that anything delivered to our door would lose its texture in the time it would take to get from the kitchen to our home.

She was, though, a great believer in the experience of eating out. For her, a delicious meal out was the ultimate pick-me-up, melting away the stress of the working week. So, over time, we three children came to learn that Friday night was restaurant night. We worked our way through all the best restaurants within a 5-mile radius of our home, and many of the worst. No stone was left unturned. From fine dining with white tablecloths and tiny portions (my least favourite) to small holes in the wall with three tables and a forty-five-minute queue to contend with (my most favourite), we sampled them all.

Our family developed friendships with restaurant owners and waiting staff, which deepened the experience for my parents, but also meant that the waiter would usually sneak an extra scoop of ice cream in each of the kids' bowls. These chefs, restaurateurs and waiting staff became our Friday night family for almost two decades, greeting us by name and welcoming us with big smiles and a funny joke for the kids.

News of good restaurants spreads fast. We came to know, from a friend of a friend who knew a guy, that a new restaurant called Spice Dragon had opened five minutes away from our home. They were serving a fusion cuisine that I had never heard of called Indo-Chinese, which seemed intriguing to me, given that I loved Indian and Chinese food in equal measure. The short drive and the prospect of ample parking was appealing to my dad, and the chance to eat two of our favourite cuisines on one plate was attractive to the rest of us. The stars had aligned in the restaurant's favour, and we were hungry, so off we went.

Spice Dragon was a tiny place, accommodating only around twenty diners at a time. The shopfront was bright orange and decorated with the image of a fierce-eyed Chinese dragon, fire spewing from its mouth and a garland of chillies adorning its neck. The tables were arranged neatly, with a series of chutneys laid out with poppadums in the centre of each one. A television was mounted against the back wall, and old Indian songs played on repeat. All four walls were painted a deep maroon colour, and large white paper lanterns hung from the ceiling. The combination gave the place a claustrophobic feeling, stifling and heavy, but the demeanour of the owner as we entered was light and welcoming. As we walked in, the smell of vinegar and soy sauce wafted through the air, making our eyes water as we acclimatised to the environment. And, as I said, we were hungry.

The menu was as intriguing as the decor. Dishes I had never heard of jumped out at me from the laminated menu: haka noodles, chicken Manchurian, Sichuan-style crispy mushrooms, vegetable lollipops, chilli and garlic paneer with green peppers. I was fascinated. As my mother ordered, the waiter, in his scarlet-coloured t-shirt, asked her how hot we would like our dishes. Each of her three children had been trained to eat food at every level of spice, so the question was an easy one for my mother. 'Medium, please,' she replied. Medium to us meant hot enough to impart flavour, but not hot enough to melt your face off.

Twenty minutes later the food arrived, looking and smelling fabulous. A discussion about the best dish of the night was a mandatory part of the journey home, so we filled our plates with a sample of each one, ready to pass judgement. Our opinions would be dissected by the rest of the family in the car, so we had to carefully consider what it was about the food that really stood out for us.

To me, the most appealing dish that evening was the chilli garlic paneer with green peppers. Cubes of crisp paneer, deep-fried in a dusting of rice flour, tossed in soy sauce and garlic and then stir-fried with thin slices of green peppers. A sprinkling of finely sliced spring onions and toasted sesame seeds adorned the mound of paneer cubes, finishing off the dish before it was served to us. Steam wafted from the serving plate into the air, fogging up my dad's glasses, but also reminding everyone at the table that this dish had only left the wok seconds earlier.

I shovelled a few pieces of paneer and green pepper into my mouth, which was filled immediately with more flavours and textures than I could recognise, each giving way to something better. Garlic and salty soy came first, followed by the earthy sesame and fresh spring onions. But when I bit down on the crisp paneer, things quickly changed. Fire spread throughout my mouth, coating my tongue and cheeks with what felt like the heat of a thousand suns. Tears immediately started streaming down my face and my cheeks flushed a deep crimson. For the next few seconds, as my body came to terms with this assault, time stood still. I looked towards my mother, her image blurry through my tear-streaked eyes. 'Mum,' I whispered, not because I was trying to be discreet, but because I physically couldn't make any sound louder than this. 'I think the paneer dish has green chillies in it.'

I was thirteen years old and, like most thirteen-year olds, I was about as self-conscious as it gets. The last thing I wanted to do was attract attention to my situation, or cause a scene, but this was too much for me to handle. By now, my nose had started watering and my lips were completely numb. I began to contemplate who should deliver the first eulogy at my funeral.

Realising how much discomfort I was in, my mother swooped in to the rescue. 'Here, drink some cold water,' she said, handing me a bottle of the cold liquid she thought would save me. I took it and drank the whole thing in one go, but by now everyone at the table had stopped eating and was watching my pink face with interest. I could see my sister smirking, trying to control her laughter. My brother let out a giggle but then, remembering that I was much bigger than him and my parents wouldn't always be there to protect him, quickly looked away, avoiding my gaze.

The water helped, but only for as long as it was in my mouth. As soon as it passed down my gullet, another wave of chilli took over. Things were progressing from bad to worse. My tongue, though painful, was not the main problem. My throat felt as if someone was inflating it with a bike pump, and my stomach was now beginning to fill with the same heat that had already made itself at home in my mouth. For the first time in my life, I experienced heartburn. Wherever the green chilli travelled, a flurry of fire and agony followed. Unexpectedly, I was also hit with a wave of adrenaline. All my senses were suddenly heightened. This was much more than a taste experience; it was real pain.

'She needs yoghurt, not water!' I heard from the table next to us. The family there had caught wind of my suffering and had clearly been in this situation before. The waiter arrived with a plateful of yoghurt drizzled with honey. My mother, looking for someone to blame for my pitiful situation, sensed her chance to intervene. 'I asked for medium, why have you given me extra hot?' she asked, not unkindly, but in a tone that only exists when a mother is telling someone off. 'You put green chilli in here instead of green peppers!'

The waiter looked confused for a moment. 'No madam,' he said. 'I assure you that this is medium.' He explained to us that the mild had two green chillies, whereas medium had four and hot had six.

Excuse me, four whats? 'Green chillies, madam.' So, a translation error with the word pepper being used as a synonym for the word chilli on the menu was the cause of my agony. There followed a long discussion between my mother and the waiter about the difference between green peppers and green chillies, during which I sat, taking sips of my honey-sweetened yoghurt, appreciating its magical soothing effects on my sweat-soaked body. I also felt very much alive, warm from head to toe, every sense and nerve ending in my body fully awake. If you squinted, I might almost have looked a bit euphoric.

As we walked out of the restaurant, I recall my mother, displaying her particular brand of humour, quipping, 'They don't call it Spice Dragon for nothing, do they?' And, near-death experiences aside, we did return on a number of occasions, usually in the winter when our craving for warming, spicy foods became overwhelming. Each time we opened the menu, we would look at how green 'peppers' had been crossed out with permanent marker and replaced with the word 'chillies' in big letters. We were, however, always careful to order mild, rather than medium. I cannot even imagine what 'hot' would be like, but I am certain I wouldn't survive it.

Since that night at the Spice Dragon, tasting small quantities of chilli and using them cautiously to flavour my food has been a hobby, but some people take things far more seriously, and nowadays, the search for the most intense heat is a competitive sport. In 2016, a man was hospitalised after attempting to eat ten chicken wings that had been coated in Carolina Reaper Chilli sauce. (He managed three.) And around the world, there are hundreds of chilli-eating competitions that pit those crazy enough to eat the spiciest foods in the world against one another. There is even a name for this subset of the culinary world; people who crave the spiciest foods are known as pyrogourmaniacs.

I am not a pyrogourmaniac by any stretch of the imagination. Instead, I prefer my foods to come with a subtle kick, the chilli a thoughtful addition to complement the other flavours. In my fridge at this moment, I can see some hot Thai red chilli paste, a jar of Gochujang spice paste, some Turkish red pepper paste, a few green finger chillies and some Scotch bonnet peppers. The store cupboard contains a selection of chilli powders of differing strengths, a bottle of sriracha sauce, some Tabasco and (because macaroni and cheese isn't the same without it) a bottle of Jamaican hot pepper sauce. Thinking about it, maybe I should reassess my pyrogourmaniac tendencies ...

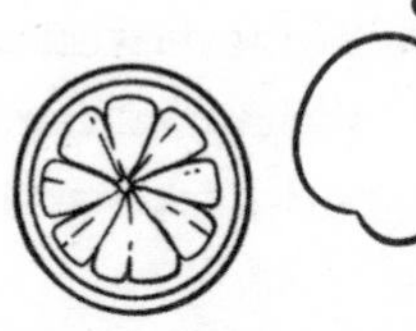

Chilli Paneer

Serves 2

This one is a proper scorcher. Add as many or as few green chillies as you can handle. Try using a chilli sauce like sriracha or Maggi Chilli Garlic Sauce (my most beloved), rather than Tabasco or hot pepper sauce, which is probably a little too acrid.

Ingredients

1 teaspoon ginger paste
1 teaspoon garlic paste
1 teaspoon soy sauce
1 teaspoon oyster sauce
1 teaspoon sesame oil
4 tablespoons chilli sauce
500g diced paneer cubes
1 tablespoon plain flour
3 tablespoons vegetable oil
1 green pepper, finely sliced
2–6 green chillies (according to taste)
1 teaspoon toasted white sesame seeds, to garnish

Method

1. Start by mixing the ginger and garlic pastes, soy and oyster sauces, sesame oil and chilli sauce together with 3 tablespoons of water. Set aside.

2. Toss the paneer cubes gently in the flour to coat them. Heat the vegetable oil in a non-stick frying pan set over a medium heat. Add the paneer cubes to the hot oil and discard the excess flour. Fry the paneer cubes until they are golden brown all over. Stir regularly while frying to ensure all sides of the cubes brown evenly. Remove the paneer from the frying pan with a slotted spoon and drain on kitchen paper.

3. While the residual oil in the pan is still hot, fry the green pepper slices in it. You want them to lose their rawness but remain crisp – this takes around 2 minutes.

4. To bring the dish together, add the paneer cubes back to the frying pan with the green peppers and pour in the chilli dressing. Toss everything together well and ensure that each paneer cube is totally coated in the chilli sauce. After 2–3 minutes in the pan the sauce will have glazed all the paneer cubes. As a last flourish, add the green chillies before serving and sprinkle with sesame seeds.

Chilli: taming an unruly beast

Nothing can take you further away from a sense of digestive health and happiness than sweating your face off because you accidentally put too much chilli in your cooking. But taming this ingredient, one so divisive that even experienced chefs might shy away from it, can create an element of mastery in your cooking – the difference between knowing three chords on a piano and being able to play Beethoven's 5th from memory.

So, what is chilli? And, how do we tame this unruly beast?

While there are hundreds of spices, there are only five domesticated chilli pepper species. But these five parent species yield thousands of completely different variants, from the sweet bell pepper to the deadly Carolina Reaper (the hottest chilli in the world, which has put many, many people in hospital). The chilli is part of the genus capsicum, and in horticultural terms is cousins with the tomato, aubergine and goji berries because of its hat-like stem (or peduncle). The word itself comes from an ancient Aztec dialect, and 'pepper' was added on when explorer Christopher Columbus and his compatriots mistook the chilli for a relative of the black pepper.

Chillies are thought to have first been eaten in 7000 BC in Mexico. The Aztecs, ever the industrialists, made a warm chocolate and chilli drink from them, inventing the spiced hot drink long before the pumpkin spiced latte was even a twinkle in Mr Starbucks' eye. But it wasn't really until the 1490s that the chilli found its way to European shores, thanks to the exploits of Christopher Columbus. From here, it spread far and wide, reaching India, Africa and eventually Southeast Asia, where many more varieties were cultivated. Seeing the chilli's potential, local communities quickly embraced the new arrival and assimilated it into their cuisines. Today, chillies are such a big part of Eastern cooking that it seems strange to even try to imagine an Indian curry or Malaysian rendang without that spicy kick.

What exactly is spiciness?

We talk about spiciness as if it is a flavour, but unlike sweet, salty, bitter, umami and sour, spiciness is *not* detected by our taste buds. Shock, horror, sacré bleu!

The family of compounds responsible for the heat of chillies are the capsaicinoids, with the best-known being capsaicin. Capsaicin is an alkaloid compound found in the flesh of chilli peppers, most concentrated in the white bits that hold the seeds in place. When you take the first slurp of your spicy laksa curry, it is the capsaicin molecules that bind to a receptor in your mouth called, rather catchily, the transient receptor potential cation channel subfamily vanilloid, member 1.

Luckily, most people who have to say this a lot call it TRPV1 (pronounced *trip vee one*) for short. TRPV1 is found in many parts of the body, not just the mouth, and anyone who has ever rubbed their eyes or wiped their bottom after handling chilli will testify to this. But mainly, the receptor is peppered over the membranes of the mouth, throat, tongue and palate. And the reason spiciness hits us so hard is because capsaicin fits around TRPV1 like the proverbial hand in the most perfect, custom-fitted glove.

The TRPV1 receptor is designed to be activated in response to extreme heat. So, when your mouth feels like it's burning after eating a chilli, it's because your brain thinks your mouth is actually, literally burning. Different people have varying numbers and distributions of TRPV1 receptors in their mouths, which explains why we feel the 'hot' sensation in different areas compared to others. From my experience in the Spice Dragon restaurant, I can tell you that my receptors seem to be situated mostly at the back of my throat, and what feels like the depths of my ears. Bizarrely, my tongue doesn't really feel the heat so much.

TRPV1 communicates with the brain via the nervous system. When capsaicin activates the TRPV1 receptor, nerves are triggered, sending a message to the brain for interpretation. But the nerves that relay messages between the TRPV1 receptors and the brain are not ones that communicate taste information, but rather they communicate touch. So, although chillies have a certain smell and many swear that they can 'taste' them, the reality is that they remain, for the most part, instruments of *touch*. And, as such, they are also instruments of pain.

But, if chillies can cause us pain, why do we go near them? No one really knows why humans started masochistically eating chillies. But not all living things are as sensitive to them; unlike mammals, birds lack TRPV1 receptors and are able to munch away happily on chillies, dispersing their seeds in the process, helping the next generation of chillies to take root.

And the truth is that we do derive some pleasure from the fiery blast of chillies. The pain they produce causes the release of feelgood endorphins, the human body's equivalent of morphine. Endorphins create a sense of euphoria, a bit like that experienced by runners who push themselves to their physical limits. Basically, when the body thinks it's about to break, it releases endorphins to make our demise a little less terrible, and this happens with traumatic experiences like chilli ingestion and exercise. The body also releases dopamine, our reward chemical, to tell us to repeat these actions, which is perhaps why some crazy people choose to eat Carolina Reaper chillies (or even crazier, go for a run) more than once in their lifetime.

I had the pleasure of meeting Shahina Waseem, an actual chilli celebrity, while filming a TV show for the BBC. She shot to international fame in the world of chilli a few years ago for eating a whopping fifty-one Carolina Reaper chillies consecutively in under fourteen minutes.

My first thought when I met Shahina was that she might carry a genetic mutation whereby her TRPV1 receptors didn't function – or at least functioned less than everyone else's. But, she explained, when she eats chillies she still experiences many of the symptoms you or I get: the red flushing cheeks, the pain, the watering eyes, heartburn (a burning sensation in the chest, behind the breastbone), as well as the inevitable 'less than ideal' defecatory action afterwards. So, perhaps for Shahina, eating a Carolina Reaper is like going on a roller coaster – she enjoys the thrill, even if the immediate sensation is one of intense pain. In fact, studies do suggest that those who like to eat very spicy foods are also more likely to enjoy other adrenaline-triggering activities like bungee jumping, paragliding or gambling.

It could be that a love of spicy food might be a genetic preference, handed down from parents to children. However, to my knowledge, no other championship-level spice queens exist in Shahina's family. But what about developing a tolerance for spices through repeated exposure to capsaicin? Can we increase our tolerance by eating more Carolina reapers? So far, studies indicate that the pain from chilli exposure probably doesn't get any better, but what *can* increase is your pain threshold. In fact, researchers have found that people who like to eat spicy foods don't experience the burn of the chili to be less agonising than those who don't, they just like to experience the agony. So, while eating lots of chillies will not make the pain any less painful, what it might do is increase your overall tolerance. Which, let's be honest, will come in handy on the toilet the next day.

This begs the question though, if chillies activate heat receptors, why do people in hot countries love cooking and eating them? Because, paradoxically, the hotter your body thinks it is, the cooler it becomes. When the heat-sensitive TRPV1 receptors in our mouths are activated, our bodies start believing that they are in contact with a dangerous heat source. This gives our brains the false impression of overheating, and the hypothalamus (an area in the brain responsible for regulating temperature) responds by activating our millions of sweat glands. The sweat then evaporates away, cooling down the body. Just watch anyone struggling with spicy food, and you can guarantee they will be sweating.

So, what can you do to stave the heat? Most often, the people around you who pity the awful state you have got yourself into will offer a glass of water. But no matter how cold the water is, as soon as you swallow, the burn will invariably return, and often feel stronger than before because of the temporary cooling effect of the water. This is because, sadly, the long hydrocarbon tail of the capsaicin molecule is not soluble in water. So, no matter how much water you drink, it won't calm that receptor down.

Milk or yoghurt, however, contain proteins that are able to envelop capsaicin molecules and, like a bouncer removing a rowdy trouble-maker from a club, hustle it out of the door by the scruff of its neck. An alternative strategy is to put something rough into your mouth, like a combination of ice and sugar, and move it around with your tongue. TRPV1 receptors can sense physical abrasion, so the mixture of ice and sugar will compete with capsaicin for the receptors' attention, and because they can't do two things at the same time you'll feel less of a burn.

Some chilli FAQs

Being a doctor and a chef, I am often asked about the health benefits of certain ingredients, and none attracts more questions than the chilli. So, I've decided to list a few of them here for you in case they were also things that you were curious about.

- **Can capsaicin make me thin?** In a nutshell, we don't know yet, but ... maybe? Evidence of capsaicin's effect on body weight is still emerging, but there are some signs that through a range of mechanisms (some of which we don't understand), chilli may be able to

Hot, hotter, hottest

We measure the spiciness of a chilli using a rating system called the Scoville Heat Index. Named after American pharmacist Wilbur Scoville, this scale was developed in 1912 and measures how much capsaicin content can be diluted in sugar solution before the spiciness is no longer detectable. A sweet bell pepper gets 0 Scoville heat units, Tabasco clocks in between 2,500 and 5,000 units and the spiciest curry available in the UK sits at around 150,000 to 300,000 units. But dwarfing all of these, the hottest chilli in the world, the Carolina Reaper, clocks up between 1,500,000 and 2,000,000 Scoville heat units.

Pepper types	Scoville heat units
Carolina reaper	1,400000 - 2,200000
Trinidad scorpion	1,200000 - 2,000000
Ghost pepper	855,000 - 1,041427
Chocolate habanero	425,000 - 577,000
Red savina habanero	350,000 - 577,000
Fatali	125,000 - 325,000
Habanero	100,000 - 350,000
Thai pepper	50,000 - 100,000
Cayenne pepper	30,000 - 50,000
Tabasco pepper	30,000 - 50,000
Serrano pepper	10,000 - 23,000
Hungarian	5,000 - 10,000
Jalapeno	2,500 - 8,000
Poblano	1,000 - 1,500
Anaheim	500 - 2500
Pepperoncini	100 - 500
Bell pepper	0

improve blood flow to various tissues, enhance energy expenditure and diminish appetite. These are all factors that help with weight loss, but sadly, studies have so far been mostly on animals.

None the less, clinical studies looking at capsaicin food supplements are currently under way, and in the future, we will know whether capsaicin has a beneficial effect in the treatment or prevention of diabetes, obesity and heart disease – conditions where the treatment involves intensive weight loss. There is also some evidence (again from animal and not human studies), that capsaicin may alter the composition of gut microbes in beneficial ways, reducing low-grade inflammation and preventing obesity. The early studies look promising, but maybe hold off on the Carolina Reaper diet for now.

› **Does capsaicin cause stomach ulcers?** In a nutshell, probably not and actually, stomach ulcers are more prevalent in food cultures where spice is used less frequently. If you already have a nasty ulcer or deep erosions in your stomach, chilli will most likely make you feel worse because of activation of TRPV1 receptors around the ulcer site, but there is very little evidence that chilli causes our stomach to auto-digest itself. In fact, most of the scientific consensus is now that capsaicin actually slows down the rate at which the stomach produces acid and stimulates the alkaline secretions that the stomach naturally produces to protect itself.

The main culprits that cause ulcers are alcohol, smoking, stress, a bacterium called Helicobacter pylori and some anti-inflammatory painkillers. If you have already developed an ulcer because of any of these, then chillies may cause worsening pain due to stimulation of the TRPV1 receptors at the site. However, the evidence for chillies directly causing ulcers remains circumstantial and, in a horrendous miscarriage of culinary justice, they have been framed for a crime they did not commit.

- **Does capsaicin make acid reflux worse?** Again, no. The notion that spicy food causes acid to rise up the gullet is based on, well, a feeling. Although spices can cause a burning, heartburn-like sensation in the oesophagus for some, this is not because of acid rising up the gullet, but the direct effect of the chilli on the TRPV1 receptors at the junction between the oesophagus and stomach. Another stitch-up for the innocent chilli.

- **Should people with IBS avoid chilli?** Well, it's complicated. The British Dietetics Association, an authority on the subject of what you should eat and when, looked at all the available evidence to do with chilli and IBS symptoms. Although they recommend a trial of chilli exclusion in people with IBS where chilli is felt to be triggering symptoms, this is based, not on clinical evidence, but on the knowledge that every case is different, and what doesn't work for one person might be effective in another. In short, evidence that chilli consumption makes IBS worse does not exist. But if you feel stopping chilli makes your IBS better, then by all means lay off the curry for a little while. (See pp. 268–271 for more about gut issues including IBS.)

Some of you may have initially thought it a bit weird to include a chapter almost entirely on spices – the least peaceful, most anarchic of ingredients – in a book on finding digestive health and happiness and harmonising your relationship with food. But I hope that the more you read, the more you'll agree that they have a place here.

I feel it is difficult to overstate just how quickly and easily a pinch of spice can elevate your food, whether you are re-inventing a classic dish or experimenting with cuisine from a new geographical region or culture. In the hands of somebody who is comfortable using spices in their cooking, these ingredients can be truly transformative, not just in individual dishes

but to their entire cooking style. Whether you are a die-hard recipe follower or a freestyle improvisor, it doesn't make a difference – spices can be used by everyone.

No ingredient can enrich your cooking like chilli, and harnessing spices will allow you to unlock the potential of food while making new 'family favourite' dishes – as long as whoever is wearing the chef's hat understands the point at which it will become overpowering. The art of balancing spices is one that takes time and practice to master. But it is these spice experiments that in some ways are the most satisfying – full of drama, colour, aroma and flavour; there is never a dull moment in the kitchen when the spice drawer is in use. In time, spice experiments will pay off, and using spices to your advantage will allow you to cook simpler, smarter and happier, while concurrently conferring some (albeit poorly defined) benefits to human health.

Building your own spice collection is a lifelong process of learning. A good starting point can be to select just a handful of spices (like cumin, coriander seeds, red chilli powder, turmeric and cinnamon) and learn to master these before widening your repertoire. I expect you will find that with practice you develop your own unique spice thumbprint, or subconscious method of balancing spices, which is as individual to you as an artist's brushstrokes.

On a personal note, for me, cooking with spice is a form of self-expression. I have stressful days when I am drawn to soothing turmeric and fennel. In contrast, if I am ever lacking in motivation, a chilli kick with cayenne pepper and heady cumin seeds seems to be what I need to awake me from my malaise. I have realised that spices can be used to mimic or influence mood, soothe, impassion, excite and comfort in equal measure – which is why using them in your cooking to find health and happiness is invaluable.

Summary

- A diverse array of spices can be found in nature, each with its own characteristic flavour profile and potential medicinal properties. Turmeric, cinnamon and ginger have been studied to a degree and seem to have some beneficial effects on human health.
- Chillies get their characteristic heat from the compound capsaicin. The intensity of a chilli can be measured on the Scoville Heat Index.
- Taste buds are not responsible for the detection of chilli. A receptor called TRPV1, which also detects extreme heat, senses chilli.
- If you have swallowed a chilli that is too hot to handle, eating yoghurt rather than drinking water or putting something abrasive like a combination of ice and sugar in the mouth will help relieve the burn.
- There is no clear evidence to suggest that chillies directly cause stomach ulcers or, indeed, acid reflux for that matter; but if you already have these conditions, they may make you feel worse.
- An exclusion of chillies from the diet can be trialled in IBS patients who feel they make their symptoms worse, although there is no convincing proof that they do.

Aloo Bun Chaat

Serves 4

Aloo Bun Chaat is an unapologetically spiced potato patty stuffed inside a soft burger bun, served with lashings of spicy green chilli and tart tamarind chutney. It is the street-food dish of dreams, lip-smackingly, tantalisingly good.

I have a particular affinity with street food; the dishes are so reflective of local tastes and seem to evoke a deep sense of belonging. Part of this arises from the use of regionally produced spices and the local cooking techniques which celebrate them, resulting in edible morsels of delight.

Ingredients

500g cooled mashed potato
2 spring onions, finely sliced
Handful of finely chopped coriander
1 teaspoon red chilli powder
½ teaspoon turmeric
1 teaspoon coarsely ground toasted cumin seeds
½ teaspoon garam masala
3 tablespoons flour
1 egg, beaten
Salt, to taste
Vegetable oil, to shallow-fry

To serve:
4 iceberg lettuce leaves, finely sliced
1 tablespoon mayonnaise
4 soft white bread or brioche rolls
4 tablespoons ready-made tamarind chutney
4 tablespoons green chutney

Method

1. Combine the mashed potato, spring onions, coriander, red chilli powder, turmeric, cumin seeds and garam masala in a bowl and season with salt, to taste. Shape into four equal-sized thick patties, the same circumference as your burger buns. Place in the fridge until you are ready to use them to ensure they remain firm enough to handle.

2. Pour the vegetable oil to a depth of 1.5cm in a non-stick pan and heat over a medium heat for around 5 minutes.

3. Place the flour in a shallow bowl and season liberally with salt. Dip the potato patties in the flour and ensure it coats all the surfaces, then dip them patties into the egg, followed by a final dip back into the flour. You are now ready to fry off the potato patties one at a time.

4. Carefully place the patties in the hot oil and fry for 2 minutes on each side, or until the surface is deep golden coloured. Try to move them as little as possible and handle them carefully with a palette knife or steel slotted turner. Drain on a plate lined with kitchen paper to absorb the extra grease.

5. Mix the lettuce leaves with the mayonnaise in a small bowl.

6. To assemble, toast the buns lightly on a griddle or non-stick frying pan. Rub a tablespoon of tamarind chutney inside the lid of each and a tablespoon of green chutney on the base. Place a potato patty on the base of each bun, and top with the mayo-coated lettuce. Then top with the lid of the burger buns and serve immediately with extra chutney on the side.

Note: you can buy shop-made green chutney or use the following recipe: blend 100g coriander and 50g mint leaves with a thumb-sized piece of ginger, a tablespoon of Greek yoghurt, 2 green chillies, a tablespoon of lemon juice and a glug of olive oil to form a smooth paste. Season with salt to taste.

Harissa Aubergines and Honey-sweetened Yoghurt

Serves 4

The name harissa comes from the Arabic verb *harasa*, which literally translates to 'pound' or 'break into small pieces'. Traditionally, it has been associated with the Maghreb region, specifically Tunisia, Morocco, Algeria and Libya. You can almost envisage the dusty, hot spice souks of Tunisia where shoppers greedily watch and wait while vendors pound chillies and oil to form vibrant pastes in front of their very eyes. Historians believe chillies were brought to Africa through the Spanish occupation of Tunisia in the 16th century. Good thing in some ways, given the fact that the use of harissa in these regions is as ubiquitous as tomato ketchup is in England.

Ingredients

Light olive oil, for frying
2 aubergines, sliced uniformly into 1cm thick rounds
¾ teaspoon turmeric
1½ heaped tablespoons rose harissa (Belazu brand, if possible)
2 teaspoons white wine vinegar
300ml thick full-fat Greek yoghurt, at room temperature
1 tablespoon honey
Handful of roughly chopped, toasted pistachios (optional)
Handful of pomegranate seeds (optional)
Handful of finely chopped parsley (optional)
Salt, to taste

To serve:
A few flatbreads of your choice

Method

1. Drizzle the olive oil liberally into a non-stick frying pan and heat over a medium heat for a couple of minutes. Toss the aubergines with the turmeric and fry in the hot oil for approximately 2–3 minutes each side, or until they soften and cook through and are slightly golden and crisp on the edges. Drain on a plate lined

with kitchen paper, which will absorb the extra oil. Be aware that aubergines do require a fair amount of oil, so you will need to top up your frying pan regularly as you prepare them in batches.

2. Once the aubergines are all fried off, toss them into a bowl and season liberally with salt. Add the rose harissa and vinegar and stir well to ensure the aubergines are all well coated.
3. Stir the yoghurt with honey and season with salt. Spread on the base of a platter and top with the aubergines. Decorate with pistachios, pomegranate seeds and parsley, if you wish. Serve with flatbreads or as a side to your favourite meat or vegetable main.

Baharat Chicken with Almond and Apricot Couscous and Dill Raita

Serves 2 greedy diners

Warm, aromatic and sweet, baharat spice mix is not dissimilar to garam masala used in the Indian subcontinent, though its roots lie in the Middle East. It often contains cumin, coriander, cinnamon, cloves, cardamom, black peppercorns, allspice, nutmeg and paprika in varying quantities: all the good stuff. You may find yourself dusting meat and fish with it, adding it to jammy roasting peppers and onions or even lacing lentil soups with it. (You can thank me later for the new inclusion to your culinary repertoire.) Here, it is used with dried apricots to make a most versatile, quick supper dish – equally fit for a lazy evening meal or entertaining guests.

Ingredients

350g minced chicken thighs
1 heaped teaspoon baharat spice mix
5 dried apricots, finely chopped
½ red onion, finely diced
Handful of finely chopped coriander
1 teaspoon salt
Vegetable oil, to shallow fry

For the couscous:

250g dry couscous

300ml salty chicken stock

20 toasted almonds, roughly chopped

10 dried apricots, finely chopped

50g chopped dill

Juice of 1 lemon

2 tablespoons olive oil

For the dill raita:

250g yoghurt (Greek is best)

20g dill, chopped

1 garlic clove, grated through a microplane

¼ teaspoon baharat spice mix

Salt, to taste

Method

1. Combine the minced chicken with the baharat spice mix, dried apricots, red onion and coriander and season with salt before mixing well to combine. Rub vegetable oil all over your hands and divide the chicken mince into eight equal portions. Take each portion and shape it into a round, flat patty, approximately 1cm thick.

2. Pour vegetable oil into a non-stick pan to around 5mm deep. Place over a medium heat before gently placing the chicken patties into the hot oil, two or three at a time. The fat should start sizzling immediately around the chicken; if it doesn't, turn the heat up. Once the chicken is golden on one side (approximately 3 minutes), flip the patties over and cook the other side in the same way. They are only a centimetre or two thick, so will cook quickly. Remove from the pan and drain on a plate lined with kitchen paper.

3. Prepare the couscous in a bowl by pouring boiling chicken stock over it and covering with a lid. Leave to stand for 10 minutes before fluffing it up with a fork. Add the remaining ingredients and mix together gently with a fork.

4. To make the dill raita, mix together the yoghurt, dill and garlic and season with salt. Sprinkle some baharat spice powder over the top.

5. Serve the dish with a layer of couscous on the bottom, and the spiced patties over the top with lashings of dill raita.

Chapter 6

HUNGER AND FULLNESS

Chefs like me might try to convince you that we eat for enjoyment, or for the taste alone, and although that is true, in a way it is also peripheral. In fact, we eat because we need to. On the most basic level, we eat because we need calories to power our bodies, and we feel hungry when we don't have enough of those. The sensation of hunger has always played a part in ensuring that our biological need for food is adequately met. We also live in a perplexing age of unimaginable food production, where agricultural bounty, obesity and industrialisation of food exist alongside malnutrition, starvation and famine.

This chapter is perhaps one of the most important in terms of our journey towards attaining digestive health and happiness, because the fact of the matter is that humans have learned, sometimes to our detriment, to eat way more or far less than we need. Some people even argue that when we talk about a person's relationship with food, what we are in fact talking about is their relationship with hunger – and the more I think about it, the more I tend to agree. How we as individuals respond to hunger and fullness determines not only our physical health, but, to a great extent, our mental health and wellbeing as well.

Societies in which people are becoming more and more ill from the effects of obesity now exist next door to ones in which millions go to sleep malnourished or starving, with devastating consequences. This is a desperate situation, whichever way you look at it. But whether it's malnourishment or obesity, hunger plays a vital role. And these two opposite ends of the hunger spectrum are not defined by geography, as you might imagine. Low-income households in more economically developed countries are increasingly likely to include malnourished children or adults, and in those very same localities there is a rise in obesity. Almost half the children in London's poorest boroughs are thought to be clinically malnourished, while at the same time more than half of the city's adult population fall into the overweight or obese categories. And the emergence of fast-food chains in developing countries has resulted in obesity in places that have little experience in dealing with such issues.

But what does it mean to be hungry? And what does it mean to be full? I think that an understanding of the phenomenon of hunger and satiety (the feeling of fullness) is vital if we want to improve, and understand our relationship with the food we eat.

Just a small side note, however, before we begin. There are many ways to define hunger, some that have to do with the availability of food or types of food (for instance, urban food deserts) and others that focus on whether people from a particular community can meet their daily minimum expected dietary requirements over the course of a year. These are extremely valuable and valid measurements, especially when tackling hunger from a global health perspective, but for the purposes of this chapter we will think about hunger in terms of the sensations we feel in our bodies: that rising dull pain in the abdomen, gnawing and worsening in intensity unless we are fed.

Now, let's get started.

Hunger in the 21st century

Our bodies are sophisticated machines, able to regulate both the amount of energy we expend and the amount we store. But humans evolved in conditions where we had to hunt for food to prevent starvation. In evolutionary terms, food supplies were either scarce or bountiful, the 'feast-or-famine' scenario. As a result, our bodies became very sensitive to negative energy balance (which is when the body takes in fewer calories than needed) and more tolerant of positive energy balance (when we take in more calories than we need). In other words, we are evolutionarily programmed to be more sensitive to feeling hungry than feeling full, and because of this, we overconsume when food is easily available, so that we can endure periods of scarcity when it's not.

Except, we aren't hunter gatherers any more, and food scarcity is now often a man-made construct rather than a natural one. People still go to sleep hungry, but nowadays that is more likely to be because their government has failed, or because there was a flood or a drought that wiped out an entire crop, or because they are pressured by their society or community to look a certain size. But globally, there is more than enough food to go around. There has been an exponential rise in the abundance of cheap, palatable food available to consumers, much of which is highly processed, low in essential nutrients, high in sugar and oil, low in fibre and often full of additives and emulsifiers. Just have a look at the ingredients and nutritional information on a packet of Oreos, Monster Munch, spaghetti hoops or processed sausages and you'll see what I mean. These foods, though they taste great, have been manipulated so much that the original ingredients involved in their production are often barely recognisable in the end product.

Imagine if a hunter gatherer stumbled into a branch of Sainsbury's. They'd be hard pushed to find the foods that they might recognise from nature, as opposed to all the crisps, biscuits, processed cereals, fizzy drinks and ready meals. They would have to make do with a waxed Granny Smith and a bag of organic chia seeds instead.

The trouble with the abundance of highly processed foods is that there is now a growing body of evidence to suggest that their high sugar and salt content makes them, well, irresistible. They taste good, so we eat them when we are hungry, but because they lack the fibre that most unprocessed food has, they don't make us feel full as quickly as they should. The more processed foods we eat, the higher the number of calories consumed and the worse off we are in terms of weight gain and the range of chronic diseases related to being overweight. And working as a doctor, believe me, I have seen what a diet made of highly processed foods alone can do to the human body.

Now, I am definitely not saying these things to pass judgement; you may recall my intense love for jam doughnuts from the introduction of this book! Rather, this chapter is designed to talk about the phenomenon of hunger and fullness from all angles, and prompt debate and discussion on the subject. Some would argue that processed foods have made it easier for people on low incomes, people who are too busy to cook or people who are both to put food on their tables and sustain themselves and their families. And I for one don't need convincing about how tasty processed foods can be – I have a four-year old son who has long since discovered chicken nuggets. But the fact remains that processed foods are so tasty because that's precisely what they are designed to be.

The science behind what we eat, and how it affects our bodies, is notoriously hard to study. Scientists can quite easily check the effect of a new blood-pressure tablet on a group of random people and a placebo on some other random people, and then compare the two. But in contrast, it is virtually impossible to randomise people for years on one diet rich in processed foods versus another totally free of any form of processed food, and map their health trajectory over time. You can't control every bit of food that enters the mouth of a study participant, and to further complicate an already complex area of research, there is really no such thing as a 'diet placebo' because everything we eat has some effect on our bodies.

So, what should we do? Brick up the entrances to all of our supermarkets? Buy a farm and only eat home-grown organic crops all year round? Of course not. I occasionally love a cheese puff and a Diet Coke, and I don't really want to consider a life without them. The answer, for me, lies in maintaining a balance between desire, food greed and hunger, and later on I will share some techniques and recipes to help manage these competing urges. But first, let's look at what hunger actually is.

The biology of hunger and fullness

As a Muslim, I normally fast during the month of Ramadan. For those of you familiar with fasting in the Islamic tradition, you will know that this is certainly not for the faint-hearted. Unlike Lent, where fasting is mercifully restricted to one particular food, in the holy month of Ramadan, Muslims fast every day between dawn and dusk with not a drop of water or crumb of food passing their lips until Iftar, the fast-breaking celebratory meal at sunset.

Like many other Muslims, I was trained to fast for Ramadan from a young age, starting with a late breakfast on the weekend when I was six or seven years old, and the following year maybe holding off on food till lunchtime. Slowly, the body learns how to master hunger cravings, and the intensity of the experience is built up until you reach puberty, when fasting for the full duration of the day becomes mandatory. There is something special in realising that a global community, made up of billions of individuals, are all feeling the same bodily sensations of hunger as you are. The aim of the fast is to put ourselves in the shoes of the many people in the world who go without food daily, and to remind ourselves of our duty of care towards those in our community who are less fortunate than ourselves. This, for me, is a deeply spiritual practice, and it goes without saying that controlling cravings, desires and pleasure-seeking behaviours makes up a big part of the process.

But for a month dedicated to food deprivation, you would be surprised at the very central role that communal eating plays. Ramadan is a month of contrasts, where deep hunger pangs rising from the pits of your stomach coincide with the most elaborate meals you will eat all year. Some of the most satisfying meals of my life have been eaten during the month of Ramadan – and I have eaten an awful lot of satisfying meals in my lifetime.

If you ask someone what they're thinking of during Ramadan, most likely they'll be thinking of food, planning their next meal, imagining the tastes and smells that they will be enjoying after sunset. Often, my mother would go into the kitchen planning on cooking one dish, and emerge having cooked five, and when I got married and saw my mother-in-law do the same thing, I wondered whether there must be something about the state of hunger that drives us to biologically fixate on food? One year, I developed a quasi-masochistic ritual where I would sit and watch the Food Network during Ramadan. I watched it every day without fail, an hour or so before dusk at the peak of my hunger. The food looked somehow better than usual, as if all of the deliciousness and colour on my screen was amplified by my hunger. As I watched Nigel Slater make homemade garlic bread or melt some cheese on screen, my stomach would churn, but I would continue watching despite the pain. The question is, why would I do that to myself?

The answer can be found in the way the body deals with hunger. In November 1944, thirty-six young men took up residence in the University of Minnesota football stadium. They were not footballers, but volunteers for a study called the Minnesota Starvation Experiment (one that probably – no, definitely – wouldn't pass an ethics committee review today). Conducted at the end of World War Two, it involved participants being asked to bring their bodies into a period of semi-starvation, where their calorific intake was approximately halved. The study aimed to answer some long-standing questions on the subject of human starvation, and was based on a laboratory simulation of severe famine. The results, which were astounding, were used to produce a guide on how famine victims in Europe and Asia should be rehabilitated. They don't make for an easy read, but do help us to understand what hunger really is.

The study showed that prolonged hunger increased depression, hysteria and extreme feelings of health anxiety (hypochondria) in many of the participants. Most of them cited extremely severe emotional distress. One subject amputated three of his own fingers with an axe but when asked why, he could not be sure if he had done so accidentally or intentionally. Sexual interest was reduced, as were strength, stamina, body temperature and heart rate. Hunger drove the men to obsess and fantasise about food. They dreamt of it, read about it, talked about it and savoured the two small meals they were given. Hunger, in almost every way, took over their psychological selves.

This study clearly represents the extreme end of the hunger spectrum, but it has allowed me to put into context some of the things I have observed during fasting, both in myself and others. The food deprivation I experience during Ramadan, though far less extreme than what the subjects in the football stadium had to deal with, make me obsess about food constantly and watch other people prepare and eat food all day. Food deprivation drove my mother, my mother-in-law and countless other home cooks throughout the world to prepare food as if they were feeding 500 when they might only be feeding five. I guess it also explains why you should never go to the supermarket hungry, because your stomach will fixate on that hunger, making sure that you fill up your trolley with food that you don't really need.

But why do I feel hungry? Why do I feel full?

We are all, I am sure, familiar with that squeezing, cramping sensation we call hunger. It was two scientists, A. L. Washburn and Walter Cannon, who did the first ever scientific research on the matter.

Washburn swallowed a balloon attached to a device which measured the contractions of his stomach. (The balloon was inflated to a certain level in his stomach and when the stomach contracted it pushed against the balloon.) Every time Washburn felt hunger pangs, he pressed a button. The study found that there was indeed a correlation between stomach contractions and hunger, and Washburn and Cannon hypothesised that the stomach contractions caused the feeling of hunger.

But as ever, the picture is more complicated. Observations of rats with their stomachs removed showed that they still experienced hunger, and humans who have had their stomachs removed (for instance, to avoid certain hereditary stomach cancers) also report feeling hungry, though less so than usual. There is clearly more to the experience of hunger than simply how shrunken the stomach is. The belly and the brain must be speaking to one another, but how?

The answer lies in how our bodies responds to three particular hormones: orexin, leptin and ghrelin. And apologies in advance here if I get a bit science-y, but I think it'll be worth it. (In reality, however, this is by no means an exhaustive account of the subject, a full review of which would be beyond the scope of this book.)

Deep in the brain lies the hypothalamus, an area about the size of a pea that controls our hunger and regulates our appetites. When a specific area of it (called the lateral hypothalamus) is stimulated, the feeling of hunger is created. A drop in our blood sugar will cause the lateral hypothalamus to release the hunger-creating hormone orexin (from *orexis*, the Greek word for appetite). There are 10–20,000 orexin-producing nerves in the human hypothalamus, which seems like a lot until we remember that there are 80–100 *billion* nerves in the brain. So, these nerves are almost always busy.

Orexin is an important appetite stimulant, and when it is released it increases our craving for food. When you miss breakfast and the time is moving close to lunchtime, but you're stuck in one of those 'Oh-just-one-more-thing' meetings, the hunger pangs that force your mind to drift away from the meeting to the comfort of a Big Mac meal are the work of your lateral hypothalamus, and of orexin. Many scientists wonder whether a drug that controls orexin production could be a future treatment to control binge eating. Studies in animals showed that when they were injected with orexin, they began eating frantically and uncontrollably, so it's pretty clear that this hormone has a massive effect on the body. However, only time and more research will tell whether blocking orexin production in our bodies is a good option for the treatment of obesity and binge eating in humans.

There is a variety of hormones secreted by various parts of the digestive system which play a vital part in our feeling of satiety or fullness. One of these is leptin (from *leptos*, meaning 'thin' in Greek). Leptin is released from the body's fat cells and sends signals to the hypothalamus, stopping us feeling hungry. Mice deficient in leptin have no hunger off-switch, and so are constantly hungry and quickly become obese. When a part of the hypothalamus that senses leptin was destroyed in mice they became dramatically obese as their brains are unable to sense fullness.

By telling our brains when we are full, leptin helps to regulate our food intake both in the short term from one meal to the next, but also in the longer term to regulate our weight. Given that leptin is secreted from fat cells, overweight people, with more fat cells, will secrete more leptin than those who are underweight. Unfortunately, one of the effects of obesity is that individuals can develop a resistance to leptin, meaning that even high levels of the hormone fail to have any effect.

Last on our list of hormones that we wouldn't want to go to the supermarket under the influence of is ghrelin. It sounds, ironically, like something you shouldn't feed after midnight (no, wait – that's gremlins!), but is actually a gut hormone, often referred to as the hunger hormone. It is produced and released mainly by the stomach, but also in the small intestine, pancreas and brain, and it stimulates appetite, increases food intake and promotes the storage of fat. When administered to humans, it can cause up to a massive 30 per cent increase in food intake, which is crazy when you think that that can mean an extra 600 calories a day on top of the recommended 2,000 for a man.

Ghrelin levels in the body are highest when we are fasting and just before we eat. When that meeting finally ends and we manage sit down with our Big Mac, ghrelin levels drop, but will rise again before our next meal. Dieting causes ghrelin levels to rise, and many researchers now see this hormone as an explanation for why diet-induced weight loss can be difficult to maintain in the long term. In a fight between willpower and ghrelin, ghrelin may often win.

Orexin in the hypothalamus and ghrelin from the stomach both make you want to eat, so scientists lump them together in a group called 'orexigenic' hormones. In contrast, leptin makes you want to stop eating and so it, along with a host of other gut hormones, are the bodies 'anorexigenic' hormones, stopping you eating (this is where the term anorexia comes from). You can picture the way our bodies control hunger as a balance: on one side you have the orexigenic hormones making you want to eat, and on the other you have the anorexigenic hormones which suppress that feeling. The equilibrium can tilt in one direction or another, promoting either weight loss or weight gain. If you strike a balance, then with a bit of luck, your weight will stay relatively stable.

Note: I have focused mostly on the physiology of hunger here, because that is where most of my knowledge lies. But the reality is that the experience of hunger has a very significant psychological dimension, too. We all have quite complicated relationships with food, and mood, emotions, stress, upbringing, emotional or physical trauma and memories all play a part in the experience of both hunger and fullness. We are all influenced by these, to some extent, but if you think that they are topics that affect you or someone you know to a degree that you find troubling, please try to address them with the help of your doctor, a qualified mental-health professional or dietician.

You're not you when you're hungry

The year that I qualified as a junior doctor I was part of an on-call rota, which meant that, for the first time in my career, I would have to face the hospital night shift. I had to manage the medical wards in the hospital in which I was working, which meant that everything, from dealing with life-or-death emergencies to writing up Gaviscon prescriptions, was in my hands.

It was three o'clock in the morning, on night-shift number one. I had spent the previous six and a half hours running around the hospital, answering countless requests for help, performing procedures and managing unwell patients. It was a scary, unfamiliar, exhausting process, but so far, I was mostly satisfied that I had things under control.

After another few hours of running frantically around, I finally found five minutes to sit down and write my notes in the patient record of the last emergency I had dealt with. The ward was dark, with just a side light for illumination. There was a heated area near my work station, and I perched on the only thing I could find, a tiny plastic stool. As I gathered my thoughts, structuring the information needed for my entry, my pen stopped working.

Now, I'm not usually someone who loses their temper over a pen. But before I knew what was happening, I became aware of a rising feeling of intense rage, directed squarely at my biro. I began jabbing the pen into the patient's notes, shaking it vigorously, becoming more and more angry at the inanimate object. The pen had failed me. How was anyone supposed to work under these conditions?

I must have let out an unconscious sound – that primal, guttural groan that comes out of your throat and tells everyone around you that you are about to blow – because the ward sister, a woman who has witnessed a thousand junior doctors have a thousand meltdowns, approached me with her own pen, offering it to me as a gift. 'Saliha, when did you last eat?' she asked, even though she already knew the answer.

I hadn't eaten since before my shift started. Nine hours of rushing around had passed since any food or water had passed my lips. 'You need to eat,' the ward sister said. 'It's not the pen's fault, you're just a bit hungry.' A box of Quality Street chocolates was promptly placed in front of me, alongside a milky cup of tea. I hunted down all the purple chocolates (the best ones) in the box and sipped on the tea, and the anger, which seconds before had consumed every fibre of my being, melted away almost as quickly as it had come on. I hadn't realised how food-deprived my body was.

The next night, I packed some snacks in the pocket of my scrubs.

I ate them at regular intervals and no further angry spells came my way. But it did make me think. Why do we experience that explosive irritation when hunger eats away at our insides? Is being hangry (hungry angry) a make-believe phenomenon, or is it based on real science? It turns out that most scientists feel that hanger is a very real thing. The term even found its way into the *Oxford English Dictionary* in 2018.

A recent study showed that low blood-sugar levels correlated with increased levels of anger and aggression in married couples. Partners were placed in separate rooms. One partner from each couple, was given headphones to wear and a voodoo doll, while their spouse was asked to blast a loud noise through the headphones. The ones with the dolls were told to stick pins into them if they felt angry with their partners. Perhaps unsurprisingly, those who had lower glucose levels stuck more pins into the dolls and bombarded their spouses with louder and longer blasts than those with higher blood-sugar levels.

But why? Because when our blood-sugar levels drop, hormones called cortisol and adrenaline are released into the bloodstream. These are our 'fight-or-flight' hormones – the same ones that would be released if you heard an unexpected noise downstairs while you were in bed. The release of these hormones affects the brain because they are related to the secretion of chemicals called neuropeptides, and the same neuropeptides that trigger feelings of hunger also trigger anger, rage and impulsive behaviours. This means that to the brain, hunger and anger look pretty similar to one another.

Hunger hacks

I suppose the million-dollar question is what we can do to avoid hunger in the first place? Well, for a start, try not to miss your meals, and eat on schedule. Include high-fibre foods (such as fruits, vegetables and wholegrains) in your diet, as well as proteins (such as paneer, tofu, beans, chicken, fish, nuts), as these will release energy slowly, preventing a slump in blood-sugar levels if you don't manage to eat for a while.

Avoid sugars (like the purple Quality Streets), as the quick rise in blood sugar will invariably be followed by a quick slump. And if none of this helps, maybe try replacing your pen more regularly on a night shift! Also, be aware that many of the products branded as 'energy bars' for control of hunger are abundant in sugar: a handful of almonds and a dried date or fig will do the trick much better than any of these products in my opinion.

While the above suggestions may seem like common sense to many of you, there are some less well understood aspects of your relationship with hunger that I will share with you in the following section. Knowing about these common 'hunger hacks' and applying them to your day-to-day life may help you in your quest for digestive health and happiness.

Can cook, will cook

Earlier in this chapter, I spoke about processed foods, and how while they have a number of very valid uses, their nutritional benefits are unfortunately lacking. Having the skill and confidence to cook a meal from scratch can significantly reduce our reliance on processed foods, which, over the course of a year, or two, or a lifetime, can make a huge difference to our diets.

And I'm definitely not talking about *MasterChef*-style gastronomy every night. Nobody does that, not even the people who have been on the show. But a repertoire of a few basic quick dishes that you can rustle up for yourself and loved ones without much fuss can really change your relationship to food. The difference between a quick homemade pasta sauce and one that comes in a jar can be huge in terms of the benefits it can bring to your diet. Cooking and eating both serve many purposes in society, bringing great richness to life, from celebration to the development of friendships, expression of religious beliefs and more. And you can't do that quite so well with a microwave lasagne. For those cooking on a tight budget, I would recommend the website and books by Jack Monroe for endless culinary inspiration.

Not all calories are created equal

Think about a piece of chocolate cake. Moist, luxurious, glistening with a thick layer of chocolate ganache, it's the quintessential forbidden food (and I'll be amazed if you aren't thinking of Bruce Bogtrotter from the movie *Matilda* right now). But is relating chocolate cake to worry and guilt, rather than pleasure and enjoyment, really beneficial to us?

A study on the associations people drew with chocolate cake, whether 'guilt' or 'celebration', was carried out by researchers who found that those who linked chocolate cake to guilt were less likely to meet their intended weight-loss goals over a period of time, showed less evidence of positive attitudes towards food and demonstrated fewer intentions to eat more healthily. If you correlated chocolate cake with guilt rather than celebration, claimed the research, you were also much more likely to feel that you lack control over your food choices.

Now obviously, both life and chocolate cake are far more complex than this study would suggest. Many of us will fall into the gap between 'guilty' and 'celebration' at times, changing our motivation around chocolate cake, depending on mood, events or any of another million things that determine our outlooks around food. However, it must be recognised that as a society, we have developed certain attitudes to food since childhood, and one of those is that eating too much chocolate is something to feel 'guilty' about. Think no further than the fate of Augustus Gloop, the chocolate-loving child in Roald Dahl's *Charlie and the Chocolate Factory*, who fell into a river of chocolate because he was a little *too* in love with it. Or indeed the aforementioned Bruce Bogtrotter, an overweight child in *Matilda*; his punishment for stealing a slice of Miss Trunchbull's chocolate treat was to eat an entire cake by himself. The cake was a punishment – something to feel guilty for eating.

But creating an association between guilt and something else is the surest way to make someone hide that behaviour behind closed doors.

Most people who feel guilty about a certain food don't stop eating it, they just eat it in secret, or feel bad after they eat it – or both.

The same researchers who studied the thinking around eating chocolate cake found that those people who associated it with celebration ate it only at those times, and therefore in moderation, and felt no guilt afterwards. Their relationships with food were consistently more positive, and they generally felt that eating the cake was something they had active control over, not something they were powerless against. And honestly, a slice of chocolate cake on someone's birthday is not going to cause any long-term damage in the vast majority of people. So, let go of that guilt and enjoy 'naughty' food in moderation.

Don't be afraid to masticate at the table

The idea that you will feel less hungry if you chew your food more rather than wolf it down in two gulps is something that has been around since the 1900s. Horace Fletcher, inventor of a concept called 'Fletcherism', basically told people to chew their food until their jaws fell off. He built up a huge following by explaining that the way to get the most nutrients from food was to chew it hundreds and hundreds of times per bite. Fletcher claimed that this process of chewing food would turn a 'pitiable glutton into an intelligent epicurean' and that 'nature would castigate those who don't masticate'. So, even if Fletcherism was a crazy fad (and it was), we now know for a fact that Horace was a master at coming up with catchy slogans.

Clearly, you don't need to chew your food a hundred times before you swallow it. It would take ages, you would bite your tongue repeatedly and you probably have better things to do than spend an hour and a half chewing your breakfast. More recent research has, however, demonstrated that there are benefits to chewing your food thoroughly. These studies show that those who chew their food more report less hunger and have a lower food intake overall. This may be because of the effect that chewing has on the various gut hormones that control hunger and appetite, but don't worry about signing up to the Horace

Fletcher newsletter just yet. The optimal chew count is currently thought to have an upper limit of around thirty-two, so you'll be done chewing a full sixty-eight chews before Fletcher and his followers.

Gather round for the feast

The term 'food commensality' is one that we don't hear very often today. It is how food researchers describe the positive social interactions that are associated with people eating together, which is most likely one of the first thing that humans did as groups. In fact, the first records (that we know of) that show humans eating together date back over 12,000, and this tradition has been consistently recorded in all cultures for many thousands of years. But they didn't have Netflix or delivery apps, so let's not judge ourselves too harshly.

Some cultures are better at food commensality than others. For instance, many European cultures, most commonly in rural villages and towns, but also in cities, value communal eating. Long tables are often seen in restaurants and homes in France, Italy, Poland and many other nations even today. The French remain dedicated to the collective ritual of communal dining, and I have French friends who simply cannot understand how we could eat a McDonald's in the car or have a sandwich in front of a computer screen. And while in the UK, restaurant-goers might order the salmon for one person, the chicken for another and the butternut squash for a third diner, further afield, in many Asian and Indian subcontinental cultures, this emphasis on individual ordering just isn't really seen; instead, the focus is on ordering food for the whole table to share. I'd argue that if there was one demographic who had found true gastronomic happiness, it would be rural Italian grandmothers making huge traditional meals for their gathered families. If only we all had an Italian nonna.

But while we can certainly learn from these various cultures, communal eating is something that has been ingrained into their relationship with food for generations, and isn't likely to take hold in busy households in the UK overnight. People in the UK and the USA are more likely than those in almost any other country in the world to eat meals in front of the television, which brings with it a number of issues. For instance, the lack of focus that we give to food while we shift our attention to the TV can make us overeat by up to 30 per cent and deprives us of what can be (if we're lucky) quality time with family.

The opposite of food commensality, though, is social isolation, a growing problem in many communities, particularly among the elderly. Here, eating in front of the TV is actively encouraged, if that TV is connected to a group of other people doing the same thing. The concept of 'digital commensality' is now emerging in some communities that don't have much in-person social interaction, allowing people who are physically isolated to connect with others via videoconferencing software. Companies such as Eatwith.com facilitate these online dinner parties, where participants prepare dishes to 'share' with their virtual guests, and can have the same social interactions as they might have if those people were inside their homes. They consistently report higher levels of overall happiness and, perhaps unsurprisingly due to the focus on the food being eaten, they also report eating less food before they feel full.

The extreme and slightly bizarre version of this phenomenon that has taken off in South Korea is an Internet activity known as 'mukbang'. Here, while you eat your dinner, you watch a 'broadcast jockey' eat huge quantities of food by themselves. There isn't really much of a social context to this activity, and it strikes me as more a form of voyeurism than anything else, but none the less, I mention it here because in a world where social isolation is growing, it's good to see how much potential digital commensality holds.

Become the boss of your hunger

The first food fad I ever heard about was when I was around eleven years old. It claimed that the physical act of eating celery burned off more calories than the celery itself contained. Therefore, or so the legend supposed, if you ate only celery you would fill your stomach while losing weight. Girls at school were eating loads of the stuff in the belief that this was how they would lose weight. (I must have missed the induction talk for this particular fad because I ate my celery dipped in full-fat cream cheese.) Now the first thing I want to say here is that I do find it worrying that at eleven years old we were so worried about our weight, and even more worrying that, as far as I can tell, this is still the case for many young people around the world.

If you're looking for digestive health and happiness in a set diet, chances are you won't find it. Diets to stave off hunger and help people lose weight can be an absolute minefield, not least because separating the minority of evidence-based methods of weight loss from the vast majority of utter stupidity can be tricky for trained scientists, let alone someone who just wants to lose a few pounds.

I remember speaking to a dietician colleague at work about the advice she gives patients when they want to lose weight. I was hoping for something amazing, perhaps a secret passed down from dietician to dietician in a mysterious ritual held once every generation, and that maybe I would be able to write about it in this book. I left disappointed, but ironically, also reassured. Her lesson for me, which has been echoed by a number of dieticians I have spoken to since, was that the real routes to weight loss are perhaps a little more boring than the general public would wish them to be. And if it doesn't rhyme with 'eat plenty of fruit and vegetables, whole-grains, nuts and seeds, try and cook your own food, drink water in abundance and booze in moderation and do some exercise', then it's probably too good to be true.

In the past, food 'experts' seemed to be in a race to find the best diet to lose weight, and often the only requirement seemed to be that each fad needed to be crazier than the one before. I mean, one fad diet that actually existed was to intentionally give yourself tapeworm, while another – one that actual humans actually followed – was to eat cotton-wool balls dipped in orange juice. Thankfully, it is now generally accepted that there isn't one foolproof strategy for weight loss, and what works for one person might not necessarily work for another (except eating cotton wool; that works for nobody).

An individualised approach that considers a person's lifestyle and food preferences is now what nearly all qualified dieticians will advise. In fact, the most up-to-date, cutting-edge research, as conducted by professor of genetic epidemiology at King's College London and author of *The Diet Myth*, Dr Tim Spector, and his colleagues suggests that we all have different metabolic responses to food. There are no 'bad' foods, only foods that are good just for 'you'. In the future, we are likely to move towards individualised nutrition, tailor-made for us, rather than promoting a one-size-fits-all approach to diet.

This chapter – indeed, this book – is not intended to be a guide on how to lose weight. My aim is to help those who want a deeper understanding of how the food they eat interacts with their bodies. If weight loss is something that you are interested in, you should absolutely seek assistance from a qualified dietician. With that said, I do have a few comments on weight loss in general, based on well-established concepts that sometimes get forgotten as we live our busy lives:

› Try to increase, wherever possible, the fibre-rich food you eat. You can add fibre to your diet by consuming more fresh fruit and vegetables and seeds and nuts, which are not only likely to make you feel fuller for longer, but will also help care for the beneficial gut bacteria that your body needs to maintain a healthy gut (more on this in the next chapter). If you can, avoid fibre supplements (in the form of powders, capsules and tablets, as opposed to fibre-rich foods) – at least for now, as the scientific evidence supporting their use to control hunger is not conclusive yet.

› Know that all carbs are not your enemy. We really do need carbs, just not quite as many as we seem to enjoy in our 21st-century diets. 'Low-' or 'free-from-' carbohydrate diets have been popular in the last two decades or so, as we know that carbs impact blood-sugar levels. Very high and very low sugar levels are linked to a number of health conditions like diabetes, heart attacks and strokes. The latest research suggests that the carbohydrate story is not quite so black and white, however. We all have different metabolic responses to carbohydrates and while some of us respond well to a very low-carb diet, others do relatively well on a higher-carbohydrate one. Even the same carbohydrate may elicit a different reaction in different people.

A great little tip, if you love potatoes, pasta, rice or any kind of starch, really, is to cool them down and reheat them once, or even twice if you have time. This increases the amount of 'resistant starches' which the enzymes that break starch down in our stomachs cannot digest. This means that you will not experience the quick spike in blood sugars that you might normally feel, instead enjoying a more gradual rise and fall in blood-sugar levels, and a more measured, steady release of insulin, all of which is useful for healthy weight management. I have included three recipes in this chapter that optimise available resistant starches in sushi rice, orzo pasta and potatoes, and can help you stave off hunger for longer.

› Be sceptical of food fads, or any colleague or friend who wants to tell you about the latest 'superfood'. Most foods are super, depending on what you want to use them for. Piling pomegranate, kiwi or goji berries on to your meals, as nutritious and wonderful as they are, is unlikely to be a golden ticket to weight loss. Similarly, lots of green tea, chilli or coffee is unlikely to help stave off hunger completely, even in supplement form. However, all of these ingredients, when enjoyed in moderation, can be very healthy parts of a wider balanced diet.

› Artificial sweeteners might have harmful effects, so avoid them where possible. There is evidence that even for those people with a normal BMI, use of sweeteners can be associated with weight gain and diabetes. Some sweeteners are also thought to affect cells deep in the brain, disrupting appetite pathways and altering the hormones involved in our process of digestion. Sweeteners will interact also with our gut microbiome in a variety of different ways that are yet to be fully understood. Plus, they taste horrid.

› Try to make time to exercise. I definitely feel hypocritical as I write this because I honestly don't even know where my trainers are any more. But deep down, we all know that exercise is a great way of expending calories and losing weight, and its benefits don't stop there. Evidence shows that exercise seems to be able to influence the body's ability to control appetite, by exerting an effect (albeit one that we don't fully understand yet) on various gut hormones that affect our drive to eat and perhaps even gut bacterial composition. So maybe it's time for me to find those shoes and relearn what a treadmill is, or just head out for a brisk walk with my neighbour's puppy.

› Sleep is an important regulator of metabolism. Inadequate sleep influences the hormones that control feelings of fullness, and can promote excessive eating. As a doctor, and mother to two boys, I have worked many night shifts and can report that when you're sleep-deprived, feelings of hunger can be amplified from manageable to overwhelming. One particular night I even recall our entire medical team being so desperate to eat, that we microwaved a whole bag of jacket potatoes just to fill our bellies. I also recall eating an entire packet of Maryland cookies after putting a weeping child to bed at 4 a.m.

But even for those that don't work night shifts or tend to children, sleep duration has decreased over the last forty years by between one and two hours in Western societies. Everyone nowadays seems to get far less than the recommended eight hours of sleep a night. From children staying up late on their tablets, to young adults out on the town or working professionals who need to be awake for long hours to bring in incomes or balance family life, the importance of sleep seems to have been lost somewhere along the way. But honestly, increasing the amount you sleep is consistently seen by professionals as one of the best ways of improving both your physical and mental health. So, what time do you call this? Get to bed.

› Stress affects all of us, irrespective of age. However, large-scale studies suggest that millennials (those born between 1980 and 1999) and generation Xers (born between 1965 and 1979) report the highest levels of stress of all age groups. As individuals, our food responses to stress can differ greatly, with some of us resorting to compulsive eating to calm frayed nerves, while others in states of stress forget to eat entirely. You will know which camp you fall into.

  But why does stress affect our weight? Levels of the hormone cortisol rise when we are stressed (which is why cortisol is called the stress hormone). If we experience this stress for extended periods, cortisol is a factor that can lead to us developing abdominal obesity (aka love handles). Excess cortisol will also make insulin more resistant, which triggers the release of our bodies' stores of sugars, which enter the bloodstream and increase appetite. Therefore, if you are stressed for whatever reason, consider addressing the causes of this stress before you attempt other weight-loss strategies, as working out how you can control it (through meditation, taking a walk, exercising, etc.) in the short term will have a positive impact on your weight in the longer term.

As I mentioned earlier, our relationship to hunger and fullness is one of the biggest roadblocks to achieving a sense of digestive health and happiness, and to improving our relationship with the food we choose to eat. However, by understanding what is going on in our bodies when we feel hunger or fullness, and by following a few simple guidelines on how to respond to those feelings, we can improve our ability to be the masters of our own hunger, and not let it control our eating behaviour. The long-term benefits for our mental and physical wellbeing of managing hunger successfully are indeed numerous.

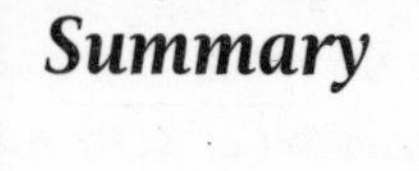

Summary

- From an evolutionary perspective, human beings are designed to be more sensitive to hunger than fullness.
- Trying to cook from scratch can reduce reliance on ultra-processed foods in the long term, with probable beneficial effects on health.
- Strategies to lose weight need to be tailored to the individual, as more research is showing us that a one-size-fits-all approach to dieting is not fruitful.
- Increasing fibre-rich foods, maximising sleep, exercising, managing stress and avoiding artificial sweeteners are all beneficial weight-control strategies.
- Chew your food properly; the number of times you chew is likely to affect how quickly you feel full.
- Multiple hormones control hunger. These include 'orexigenic' hormones that make us want to eat and 'anorexigenic' hormones that stop us. The balance can tilt in one direction or the other.
- There is a psychological dimension to hunger that accompanies the physiological.

Double Cook and Cool 'No-Guilt' Potato Salad

Serves 4

I love a hot spud. But there is really no denying the fact that cooking potatoes, followed by cooling, then reheating and finally cooling them again before consumption will increase the number of 'resistant starches' (see p. 181). So, well worth a try instead of a jacket potato.

Ingredients

600g boiled new potatoes
300g Greek yoghurt
3 garlic cloves, minced
30g chopped dill
50g cornichons or gherkins, roughly chopped
1 teaspoon honey
1 teaspoon red chilli flakes
Juice of 1 lemon
2 tablespoons olive oil
Salt, to taste

Method

1. Halve the boiled new potatoes and allow them to cool completely. Once cool, reheat them in the microwave for around a minute, and once again set aside to cool. I always keep the skin on the potatoes as it is a fantastic source of fibre.

2. Mix the Greek yoghurt, garlic, dill, cornichons or gherkins, honey, chilli flakes, lemon juice, olive oil and salt together in a large bowl to form the dressing for the potatoes. Tip the potatoes into the dressing and mix well to ensure each potato is coated fully.

3. You can serve there and then, or alternatively leave them in the fridge until you are ready to eat them. I particularly like eating this with smoked fish, like salmon or mackerel or a slice of roast beef for Sunday lunch.

Orzo with Orange and Roasted Peppers

Serves 4–6

A practical dish for lunch, or a lunch box. Again, the process of cooking and cooling the orzo will help develop resistant starches on the pasta. You could just as well use other pasta shapes of your choice, or wild rice that has been cooked and cooled twice over. (When reheating rice, always check that it is steaming hot all the way through.)

Ingredients

6 red or yellow peppers, halved and deseeded
5 tablespoons olive oil
250g orzo
3 large oranges
Juice of 1 lemon
1 teaspoon sumac
50g parsley, chopped
2 tablespoons rose harissa
Handful of chopped roasted hazelnuts (optional)
Handful of pomegranate seeds (optional)
Salt, to taste

Method

1. Preheat the oven to 180°C fan. Place the peppers on a baking sheet, drizzle over 2 tablespoons of olive oil and season liberally with salt. Transfer to the oven and roast for approximately 30 minutes until the peppers are softened completely and charred on the edges. Remove from the oven and set aside to cool.

2. Meanwhile, boil the orzo in heavily salted water, according to the packet instructions. It usually takes about 5 minutes. Drain the pasta and allow it to cool completely.

3. Slice off the top and bottom of each orange. Next, use even, downward strokes to slice the peel away from the flesh. Discard the peel. Now cut between the membranes to segment the pieces of orange flesh. Once you have done this, the remaining pith

holds a fair bit of orange juice, which you can squeeze out and keep aside instead of wasting it.

4. To assemble the salad, place the cold orzo in the microwave for a minute, then transfer to a large bowl. Add the orange segments and juice followed by the roasted peppers, 3 tablespoons of olive oil, lemon juice, sumac, parsley and rose harissa. Give everything a thorough mix and taste. Add some salt if you feel it needs more, which it probably will, then give one final mix before spreading the salad on a large serving platter. Top with hazelnuts and pomegranate seeds as a garnish, if desired, and serve.

Steak and Sushi Rice with Pickled Ginger, Sesame and Seaweed

Serves 2 generously

The cooled sushi rice is rich in resistant starches, making you feel full for longer: not much sushi rice is needed to fill you up. I love using sushi rice as it acts like a sponge for strong flavours – here ginger, soy and sesame with a kick of chilli and seaweed are the stars. You can replace the steak with very fresh salmon or tuna, or even tofu or slices of stir-fried mushrooms, if you are vegetarian.

Ingredients

125g uncooked sushi rice
2 x 250g sirloin steaks (approximately 2cm thick)
1 tablespoon vegetable oil
½ cucumber, deseeded and sliced into long, thin matchsticks
4 spring onions, cut into fine shreds
2 tablespoons pickled sushi ginger
Sea salt flakes

For the dressing:
4 tablespoons dark soy sauce
2 tablespoons rice wine vinegar
1 teaspoon sesame oil
½ teaspoon wasabi paste (more, if you like it really hot)

For the spice powder:
1 teaspoon sesame seeds (white or black)
1 teaspoon aonori seaweed flakes or nori seaweed flakes
1 teaspoon red chilli flakes

Method

1. Place the sushi rice in a pan and prepare it according to packet instructions. This usually involves topping the rice with water, allowing it to come to a rolling boil before turning the heat down and placing a lid on your pan. The rice tends to be simmered for 10–12 minutes until all the water has been absorbed before being set aside to cool.

2. Meanwhile, place a heavy-duty, thick-based, non-stick pan over a high heat until hot (but not quite smoking) and place the steaks in the pan. Season generously with salt and drizzle over the vegetable oil. Try not to move the steaks too much as you want to give them time to char. Turn after around 2½ minutes and allow the other side to cook for the same length of time (and char). This should leave you with steaks that are cooked medium. Leave the steaks to rest on a plate, covered with foil.

3. To make the dressing, mix the soy sauce, rice wine vinegar, sesame oil and wasabi in a small bowl and stir well to combine. To make the spice powder, place the sesame seeds, seaweed flakes and chilli flakes in a mortar and pestle and give them a good grind to combine and form a sort of coarse powder.

4. To serve, spread the sushi rice on a platter and top with the cucumber, spring onions, and pickled ginger. Cut the steaks into thin slices and scatter, along with any remaining juice, over the sushi rice. Spoon over the dressing and sprinkle over the spice rub before serving.

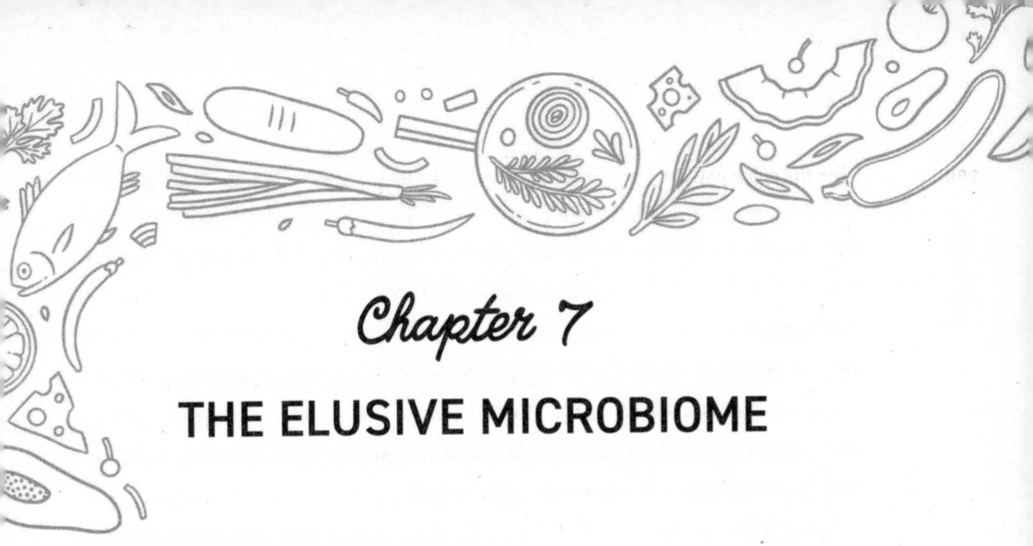

Chapter 7

THE ELUSIVE MICROBIOME

When I gave birth to my first baby, I damaged my pelvic floor (the group of muscles and ligaments that support the bladder, womb and bowel) pretty severely. I'm 5 feet 2 inches tall, with a size 6 frame, weighing about 50 kilograms, so it was a bit of an ordeal (i.e. seventy-eight hours of indescribable agony) to push out my gorgeous but chunky 9-pound baby boy. For my sins, I suffered the not-uncommon side effects of natural childbirth: immediately following the birth, I had horrendous haemorrhoids and a nasty third-degree tear, along with longer-term effects such as a weak bladder and impaired bowel function.

The whole process was enough to put me off childbirth for five years. No matter how cute he turns out to be, having a 9-pound melon bulldoze its way through your private parts can leave a person both mentally and physically scarred. Towards the end of my second pregnancy, I mentioned these fears to my midwife, who convinced me to meet an obstetrician. The obstetrician, a lovely woman with years of experience, recommended that due to my weak pelvic floor and the trauma of the first birth, I should have a Caesarean section. I came home that evening and recapped the events of the day with my husband. An unexpected tear rolled down my cheek as I told him that I had been listed for an elective C-section in a fortnight.

My deflation was because I felt like the chance to bring my second child to the world naturally was being taken from me, a feeling that I am sure will be familiar to many women. With all the compassion in the world, my husband did his best to console me. He covered the usual bases: that a C-section was an incredibly safe procedure with low rates of complication, that he would be there to care for me afterwards, that the people operating on me would be very experienced and that I would be in the safest hands there were. Then he asked, 'Are you worried that you won't have the same bond with our number two because you won't be giving birth to him naturally?'

In reality, it was none of these things. 'Then what's worrying you?' my perplexed husband queried. 'You don't want all those pelvic-floor problems you had last time again, do you?' And yes, that did bother me, but strangely, my most pressing concern was for my baby's gut. You see, when you deliver vaginally, the baby is introduced to a world of beneficial natural bacteria as he travels down the vaginal canal. A C-section, in contrast, is a sterile procedure which, by design, is devoid of any bacteria, beneficial or not.

So, in the most attractive words to ever leave any woman's lips, I replied, 'I'm worried about not exposing the baby to my vaginal flora!' And I went on, distraught: 'He won't develop his gut microbiome properly and it is going to affect him for the rest of his life!' My husband looked at me with a mixture of confusion and amusement. He knew that as a gastroenterologist I was obsessed with people's gut bugs and bowels, but hadn't, until that second, fully appreciated how much I cared about the gut of our unborn child. Childbirth is the first time that a child's previously sterile nose and mouth are colonised by their mother's bacteria. It's the start of a lifelong relationship with bodily microbes (very small living things that can only be seen with a microscope), and a C-section would mean that my second-born wouldn't have the best chance to be exposed to the 'good bacteria' that the gut needs to work properly, the way nature intended.

I did end up having an elective Caesarean section to save my pelvic floor. I weighed up the pros and cons, and concluded that a C-section was a fair price to pay for being able to hold my wee in. But, even today, months after the delivery of our second, perfectly healthy baby boy, I still worry about the impact that my decision might eventually have on his gut (no, really, I do).

Now here's a question for you: how many views does the video 'How bacteria rule over your body – the microbiome' have on YouTube? It sounds pretty niche, so maybe a thousand? Five thousand? A hundred thousand? Would it surprise you to know that the answer is nearly 6.5 *million*?

The gut, once dismissed as uninteresting by researchers, is enjoying a renaissance thanks to some very interesting breakthroughs in our understanding of how the health of this area of the body can impact our lives. It has captured the interest and imaginations of the general public, and newspaper headlines like 'Keeping your gut microbiome happy is the key to healthy eating' and 'How moving to another country could upset your gut bacteria' (both actual recent headlines from UK newspapers) have become increasingly common.

Your Microbiome: the Organ You Never Knew Existed

The magical interplay between human beings and the bugs that live inside us is at work from cradle to grave and is, therefore, well worth our time and understanding.

If you believe, like almost all of us do, that your body is made up of your own cells possessing your own individual DNA, think again. Your body is, in fact, a vessel, hosting a range of specialised microbes. From the Greek word 'bios', meaning life, these microbes live everywhere – on your skin, in your mouth, in your vagina (if you possess one) and most abundantly in your digestive system. Some scientists estimate that over the course of a lifetime, a human will play host to bacteria that collectively weigh the equivalent of five African elephants. I am not sure how exactly they came to make these calculations, but the point is clear: the microbiome is not an insignificant entity.

The human gut hosts around 40 trillion microbes. For those needing to conceptualise that number, it's a four followed by thirteen zeros; it outnumbers the stars in the Milky Way (a mere 300 billion), all the fish in all the oceans on earth (around 3.5 trillion) and even the individual cells in your body (around 30 trillion). Thirty trillion human cells and 40 trillion bacteria is almost a one-to-one ratio of human cells to microbes, and this gargantuan mass of microbes weighs around 2kg. Because it would be a pain to remember the names of all these microbes, we refer to them collectively as our gut's microbiota.

The human body forms a landscape of microbial habitats that is more diverse than any landscape on earth, and each individual's microbiota is unique – almost like our very own personal fingerprint. According to the findings of the Human Genome Project, humans are made of just shy of 21,000 genes, which is around a thousand more than an earthworm. The genes that make up the microbiota in the human body is termed the 'microbiome'.

So, to recap, the microbiota are the bugs, and the microbiome is the collection of genes inside all those bugs put together.

When researchers analysed the human microbiome, they found that it contains a *huge* number of genes, literally millions and millions. This far outstrips the 21,000 human genes that make our bodies what they are, but these millions of genes don't stay the same for ever. The composition of the microbiome is unstable, susceptible to rapid and frequent changes, unlike our human genome. This is a good thing, because it means that when we change our diets – for instance, we switch to the paleo or the keto diet – our microbiome is able to adjust. It is also able to adjust if we find ourselves in a state of starvation.

When a baby is born and passes through the vaginal canal, their nose, mouth and skin are colonised with their mother's flora, and this is the point at which the gut microbiome begins to develop. Some recent studies have shown there may even be microbes in meconium and traces of bacterial DNA in the placenta and amniotic fluid, raising the question of whether the microbiota of the foetus begins to develop in the womb itself. Their first food, breast milk, helps the gut microbiome to mature, and as the child grows, the microbiota matures with them until they are around three years old, by which time a diverse array of bugs will have colonised him or her all the way from their mouth to their bum.

As you journey through the bowel you find that the composition of the microbiome varies according to where you look. If you looked into the highly acidic stomach, you probably wouldn't find much microbial life at all (though it does exist), but as you move down the small intestine, this number grows to around 10,000 microbes per millilitre of gut content. It's only when you get to the junction between the small and the large intestine that the bigger populations of gut bacteria are evident, with approximately 10,000,000 microbes per millilitre here and in the large intestine. And deep in the dark recesses of the colon, we find the bacterial metropolis, which around 10^{11}–10^{12} microbes per millilitre of gut fluid call home.

OK, so now that I have given you an idea of quite how important bacteria are to our bodies, and how much of our bodies are just bacteria, we can start talking about what they actually do. For instance, when you watch an advert for one of those yoghurts that say they contain 'good' bacteria, what does that actually mean? Well, there are three main categories of 'good' bacteria (Bacteroidetes, Firmicutes and Prevotella actinobacteria, in case you are interested), and we all possess different strains of these three groups (among other bacteria) in different ratios, as well as a range of fungi and viruses.

It's actually quite simple to get the composition of your microbiome tested. All it takes is a Google search and a quick poo in a cup, which you then send off for analysis to one of a number of established companies. Generally, the more diverse your microbiome is, on a scale called the Simpson index, the better. I have thought about sending my own sample off, but always chicken out at the last minute because of the potential embarrassment of a gastroenterologist having a bad assortment of colonic bugs (I joke, but actually not). Perhaps one day

I'll pluck up the courage to put myself in the spotlight. A number of celebrities and popular scientists have done it, and as a result, it's becoming increasingly popular to know what makes your gut tick and how you can make it work better.

All living creatures need to be fed, and your gut bacteria are no different. So, when you eat an apple, who is it that you are actually feeding? Well, the apple you ate in the hope that it would keep me and my colleagues away is feeding both your own body cells, and also the bugs in your belly. Once it passes through your belly, the microbes in your colon digest the bits of it that the small intestine isn't able to break down. The non-digestible bits are fermented by the colon's microbes, yielding energy that could not otherwise have been harvested, recycling nitrogen, sugars and fats that escaped digestion in the small bowel and even assisting in the production of vitamins B and K.

Gut microbes have co-evolved with us over time. They love us because our guts provide the perfect warm, wet environment for them to flourish, and over millions of years of human evolution, they have become an essential part of our digestive process. What's more, their help is not limited to digestion; the microbes of the large bowel have been shown to influence our levels of immunity, and also to affect our bodies' ability to keep inflammation contained, stimulate the local nervous system and help increase the beneficial cell turnover on the lining of the gut wall. They're just so talented.

A beautiful, balanced equilibrium exists, therefore, between our bodies and our microbiota, but there can often be collateral damage in our fight against other ailments. For instance, a course of antibiotics or the explosive laxatives required for a colonoscopy can practically wipe out our gut microbiota. However, we now know that the appendix – an organ that for years we thought was useless – harbours a reservoir of microbes which can start the process of colonising the gut once again, restoring our natural environment. Bad news for anyone who had their appendix removed before scientists figured this out, but good news for the rest of us if we ever need to take a course of strong antibiotics.

With all this said, the point I am making is this: a vital step towards achieving digestive health and happiness lies in getting to grips with the concept that you are host to a vast array of gut bugs that require care and nourishment, just like every other part of your body. The steps on how you can do this through the food you eat will be addressed later in this chapter.

The microbiome and obesity

As a medical student at King's College London, I used to take long bus journeys deep into south London for my placements. One drizzly, grey autumn day – the sort that cottage pie and a warm fireplace are made for – I overheard a particularly memorable conversation between two schoolgirls, probably around fifteen or sixteen years old.

They were having a very serious-sounding discussion about what contributes to obesity, while sharing a pack of Haribo gummy bears. 'I really think that we need to be able to talk about body weight and shape without getting worried we will offend people,' said the first. 'It's not about body shaming, it's about having a conversation. If we don't talk about what makes people gain weight, then we can't reverse it.'

The second girl agreed, and then jumped right into the big question of the day: was obesity genetic? 'I think it's mostly to do with your genes, and body weight runs in families,' she mused. 'On my mum's side all the women are big and none of us eat that much. But the women on my dad's side all look like sticks and they eat like bloody horses.'

They both threw this around a little bit, while working their way through the Haribos. 'I think that a lot of it is to do with your metabolism,' said Girl A said, hitting on another of the big questions posed in the field of obesity research. 'Maybe your dad's side just have really fast metabolisms?'

I'm not sure what the two girls concluded, as they got off the bus a few stops before I did, along with their empty Haribo packet. But their conversation touched on a number of the issues that many researchers are tackling as a matter of urgency. Obesity is fast becoming a public-health crisis, and its causes aren't as simple as 'if you eat too much, you'll become obese'. As my two new friends correctly identified, sometimes people who eat less than others can gain more weight. But why is that? The answer, as you will have guessed (given that the story is in this chapter), lies in our microbiome.

When speaking to patients in clinic about their weight, the conversation between those two schoolgirls often comes to mind. It reveals how many of us frame the causes of obesity in our minds as a combination of too many calories, too much takeaway, genetic predisposition and this idea that some people are just 'lucky' with a quick metabolism, while others aren't. However, the truth isn't even close to being that simple.

Do genes contribute to obesity?

Genes do seem to have a little to do with weight gain, but it cannot be overstated that genetics is only one factor in an extremely complicated issue, and cannot alone explain obesity, or why so many people are overweight these days. I say 'these days' because only a century ago, rates of obesity were much lower than they are today, and in evolutionary terms a hundred years isn't even a rounding error, definitely too short a time for any sort of natural selection to shift us towards obesity. Typically, around a hundred generations, or 2,500 years, are needed before any genetic adaptations even become noticeable.

Very rarely, a gene called MC4R is defective, resulting in compulsive overeating and, therefore, obesity. But in most people, no single obesity gene is at fault. Genome-wide association studies have found around fifty genes to be associated with obesity, and given that the human genome contains 21,000 genes, fifty is not really anything to write home about.

So, genes contribute, but are not the whole picture. What about this thing called metabolism?

Does a low metabolic rate make people fat?

The notion that our metabolism (or to use the scientific term, our basal metabolic rate) affects weight is probably a fallacy. In fact, those of us who have higher body weights actually have higher metabolic rates than those who weigh less, as the larger a body is, the harder it has to work to produce more energy to feed all of its cells. Putting aside the small number of cases where there is a dysfunction of the thyroid gland, the claim that fast or slow metabolism affects weight is simply not true.

Are people overweight simply because of the number of calories they consume?

The widely held notion, that obesity is the cut-and-dried result of calorie intake versus energy expenditure is increasingly being questioned by research. Of course, calorie excess has a role to play, but it does not explain the fact that two people can eat the same number of calories and be dramatically different body shapes.

So, why can two people can eat the same number of calories but have dramatically different body shapes? If you think my answer is going to contain the word microbiome, then you've definitely been paying attention.

Can my gut bugs make me fat?

In a word, probably. There is a growing body of evidence to say that the gut microbiome plays a big role in obesity. The most compelling comes from studies carried out on mice.

The link between obesity and the gut microbiota was initially suggested on the basis of studies on 'germ-free' mice, raised in sterile conditions and therefore lacking a developed gut microbiome. When the microbiota from the guts of normal mice were transplanted into these germ-free ones, some surprising findings came to light. The previously germ-free mice showed up to a 60 per cent increase in body fat within two weeks, without any increase in the amount of food they consumed.

Moreover, when obese mice have had their microbiota compared with those of lean mice, noteworthy differences in the composition of the microbiota have been observed. Obese mice were found to have higher proportions of Firmicutes and fewer Bacteroidetes; this is a combination that favours weight gain, as Firmicutes extract a greater amount of energy from the food you eat compared to other bugs, resulting in more calories being released and entering the body. Similarly, when human studies were undertaken, it was found that obese individuals were shown to have more Firmicutes than leaner ones.

To put it another way, If Bob, who consumes 2,000 calories a day, has a microbiome that is more Firmicute-heavy and therefore able to extract, say, 2 per cent more calories from food than Jeff's, then Bob will extract 40 extra calories from what he eats each and every day, even if he eats exactly the same amount of food (containing the same number of calories). Now, 40 calories may not seem a lot (it's around half the calories in a digestive biscuit), but over the years and decades this might translate to weight gain.

Coming back to the point I made at the start of this chapter, studies have shown that babies who are born by C-section (and hence don't kickstart their microbiome with healthy microbes from the vaginal canal) are far more likely to be obese children and overweight adults. This is the case even if they are brought up in the same household as brothers and sisters who were born naturally, and they all eat the same food growing up. However, the literature seems to suggest that breast-feeding is very advantageous in terms of nourishing the microbiota and can (hopefully) compensate for the C-section to some extent. Come back to me in eighteen years and I'll let you know how we got on.

The curious link between blood sugar and gut microbes

We associate cake with obesity, with being overweight, with overindulgence. But why? Well, it's all about the sugars. When you chew and swallow that piece of chocolate cake, it passes through the stomach into the small intestine, where a variety of digestive enzymes secreted by the pancreas break down the long carbohydrates of the cake into shorter, more useful sugars. Now, the surface of the small intestine isn't flat, but is lined with finger-like projections called villi. These, and the microvilli that live on them, increase the surface area available for you to absorb nutrients almost *600-fold*. It's an incredibly efficient organ which, through evolution, is now highly specialised in extracting huge amounts of nutrition from your food.

The absorbed sugars enter the bloodstream giving a spike in recorded blood-sugar levels, but it appears that your microbiota is once again the puppet-master in this process. Current research suggests that the particular ratio and composition of your gut bacteria are able to influence how much your blood sugar responds to certain foods. The PREDICT study has gone as far as identifying fifteen 'good' and fifteen 'bad' gut microbes that are linked to better or worse heath, including blood sugar control, inflammation and weight. For example, a microbiome rich in *Prevotella copri* and *Blastocystis* species is associated with healthier blood sugar responses after eating a meal.

So, because of our microbiome, we are all different in how much or how little our blood-sugar level will spike in response to that piece of cake we eat. The higher the sugar spikes, the harder the pancreas has to work to secrete insulin and the greater the chance of gaining weight or developing Type-2 diabetes. Given the highly personalised composition of each individuals' microbiome, the research implies that we may be able to modify our gut microbiome to optimise our health by choosing the best foods suited to our unique biology. If you take a step back to think about this, this is HUGE. If your microbiome is, as now seems to be the case, responsible for the huge variances in how different people process sugars and other nutrients, it will be interesting to see how the field of weight management evolves over the coming years. Will we still focus on 'calories in, calories out'? Will we still shame people for the occasional takeaway?

I hope that in years to come, health professionals will provide people with individualised plans to assist in weight loss, focusing on gut health as a key contributor to the success or failure of any weight-management plan. We should be told which foods cause those worrying sugar highs so that we can avoid them, and which allow our bodies to maintain a steady sugar state for longer. We will probably be more likely to be in a position where we can advise people on the particular probiotic supplements they would benefit from to create a healthier, more diverse microbiome that favours a lean body type.

A final note on this topic: research has shown that a procedure called a faecal transplant (whereby a healthy, lean person's poo is endoscopically transplanted into the digestive tract of an obese individual) has shown potential to reduce obesity levels significantly. The work is still in its infancy and is not part of standard clinical care at present, but the world of obesity research and management is going to get very exciting over the years to come, with burgeoning understanding of the gut microbiome.

Food and the microbiome

By now, you might be wondering what all your new-found scientific knowledge about the microbiota has to do with a book called *Foodology*, or with finding digestive health and happiness. Well, the bottom line is that the food we eat has a direct influence on the composition of our gut microbiota, and once we realise the importance of our guts' inhabitants we can begin to make food choices that help them to work to their full potential.

Current research suggests that the gut microbiota is influenced by diet, and that some microbiota are 'better' and more diverse than others. A healthy, diverse microbiota has been linked with a variety of health benefits, like lower risk of heart disease and diabetes, obesity and metabolic disease, among others.

Research now seems to show that people who exist on a Westernised diet (pizzas, tacos, that sort of thing) can lose up to *a third* of their microbial diversity. Fifteen thousand years ago, people would eat more than 150 ingredients per week, whereas most of us now eat fewer than twenty. In fact, of the 250,000 known edible species of plant on earth, we use fewer than 200, with three-quarters of food that we consume globally coming from just five animal and twelve plant species. How dull are we?

Optimising gut health naturally

One way of diversifying the microbiota is to introduce a greater variety of foods into our diets, particularly those that are plant based. Now, I am not advocating switching to an exclusively vegan or vegetarian diet, but instead encouraging (gently, I hope) the introduction of a greater assortment of vegetables into your existing regime. The judgement call on whether to include meat or not is one that everyone needs to make themselves, based on their own lifestyle, beliefs and taste preferences. But it is a fact that eating more plant-based foods will make a positive difference to your microbiota, and relatively rapidly. Studies conducted at Harvard University assessed the microbiota of healthy volunteers who were switched to either a plant or meat-based diet and found that the microbial composition of both groups adapted rapidly to the shift.

Like any dietary change, adding more plant-based foods to your diet is not necessarily an easy, or an economical task. In nearly any supermarket, the aisles selling plant-based foods are far outnumbered by those selling cheaper processed foods. And given what we know about hunger from the previous chapter, never go into a supermarket on an empty belly, unless you want to come out with a trolley full of crisps, cookies and croissants.

Cooking plants is an art form. They command a certain respect, but nobody is good at something immediately, and so I fully encourage you to experiment and learn how to handle plants in the kitchen. The first milestone is knowing how to avoid creating a waterlogged, tasteless mess, which I will be the first to say is no mean feat. Once you have this skill down, you can then experiment with the hundreds of ways to optimise the flavour and texture of your plants.

Clearly, there is a vast difference between soggy canteen-style mushy broccoli, and Yotam Ottolenghi's char-grilled sprouting broccoli with sweet tahini; the two evoke two vastly different responses, both in appetite and emotion. Cooking plant-based dishes can take you on exotic culinary journeys around the globe, while at the same time feeding the zoo of microbes within you, and this makes cooking more vegetables a great hobby to get good at.

I have found that an economic way of adding plant-based foods to your diet on a budget is to ransack the freezer aisle at your local supermarket. And smaller Middle Eastern and Indian food shops might even have those exotic-looking frozen vegetables that you can't find at Sainsbury's. Some people say that freezing vegetables removes some nutrients, but this is wrong; it only affects their nutritional composition minimally (if at all), despite the bad press that they have received in the past compared to fresh vegetables. I have bought frozen okra, cassava, bitter gourd, mustard leaves, Persian molokhia leaves, cranberries, broad beans and so much more, and they all taste wonderful and have become an indispensable part of our family diet. For most of us, our repertoire of frozen vegetable use stops at the bag of peas that has been gathering frost in the back of the freezer for the last two years, but the potential of the freezer section to improve our gut health is huge, and I encourage you to take advantage of it.

Signing up to a vegetable box delivery every week is a great (albeit less economical) way of forcing you to try out vegetables that you haven't had before, and they are often accompanied by inventive recipe cards. The last few years have also seen a huge rise in the availability of vegetarian and vegan cookery books, catering for the growing number of cooks who want to experiment with more interesting plant-based dishes. I have learnt myself to incorporate more vegetables in my cooking in simple ways; Monday in our house is now 'Meat-Free Monday' and Wednesday is 'Vegnesday', and both have gone down well with my family. The jury is still out as to the success of 'Tuber Tuesday' and 'Brassica Bank Holiday'.

Cultivating your gut microbiota

Another way of cultivating the microbiota is to treat your gut to a diet rich in prebiotic and probiotic foods.

A beginner's guide to prebiotics

Prebiotics are foods that are known to promote the growth of healthy gut bacteria, and can include wholegrains, apples, bananas, leeks, asparagus, dandelion greens, cauliflower, broccoli, Jerusalem artichokes, chicory, honey, garlic, seeds, nuts, lentils, cocoa and green-tea extracts. You can think of them as being a bit like the Miracle-Gro you apply to your flower bed – they fertilise the bowel and encourage the growth and proliferation of the microbiota in your gut. I almost always have a Tupperware box full of my Prebiotic Tabbouleh in the fridge, it is my go-to gut-cultivating salad, nutritious and delicious in equal measure (see p. 217). Prebiotics contain fibre, which we would be unable to digest without our microbiota. The microbes break fibre down to release a chemical called butyrate, which plays a key role in keeping the gut wall healthy and helping to maintain its barrier function, as well as decreasing inflammation.

Chicory is particularly high in a dietary fibre called inulin (not to be confused with insulin, which regulates blood sugar and is released by the pancreas). Inulin has been shown to play a role in the prevention of heart disease and the promotion of good bone health. Many people are perplexed at how to cook chicory, owing to its bitter undernotes, but I find pairing it with sweet fruit, like apricots, grapes or peaches is a fantastic way of making this nutritious vegetable more palatable. If you are a chicory novice, I would highly recommend trying my Apricot, Roquefort and Chicory Salad with Grapes, Candied Walnuts, Chilli and Mint (see p. 218). It is summer on a plate.

A whistle-stop tour of probiotics

In contrast to prebiotic foods, probiotics are foods that actually contain live bacteria. The intention is that these bacteria find their way into the bowel, where they can make their home and lots of bacteria babies. The theory is that eaten in the right quantities, probiotics really can help the gut microbiome stay healthy. And I say theory, because for many probiotic foods there are simply not enough studies confirming whether they contain enough beneficial bacteria to be deemed effective.

The other challenge is that not everyone finds these probiotic foods instantly delicious the first time they try them, which is where the work of us chefs comes in. For the most part, they are produced by fermentation (more on this later) and therefore have a somewhat pungent smell and lactic appeal that can be an acquired taste. I would not worry if you don't like various probiotics at first, but do keep trying them as they seem to have a funny way of growing on you.

Here are some popular probiotic products (this is by no means an exhaustive list):

- **Sauerkraut:** made from shredded raw cabbage fermented in brine by various lactic-acid bacteria. Varieties of sauerkraut exist across Eastern and Central Europe, but also other countries like the Netherlands, where it is known as 'zuurkool' and France, where it is called 'choucroute'. In Germany, it is traditionally eaten on New Year's Eve as a sign of good luck (it is believed that the amount of wealth to be gained in the coming year is proportionate to the number of sauerkraut shreds consumed). Sauerkraut has also made its way across the pond to America. It was actually in an American soul-food cookery book that I came across a wonderful recipe for a sauerkraut and pineapple salad, which I have adapted for you to enjoy (see p. 220). It is a rather exciting way of using sauerkraut, rather than serving it as a side to meat or as a sandwich filling.

  To make your own sauerkraut, finely shred 2 kilograms of

cabbage (red, white or both) into a large mixing bowl. Rub about 5 heaped tablespoons of flaky sea salt into the cabbage. With clean hands, keep rubbing in the salt for around five minutes, until the cabbage has reduced quite a bit in size and released a fair bit of liquid to the bottom of your bowl. You can add spices of your choice at this point. Options include caraway seeds, juniper berries, pink peppercorns, apple, fennel or even garlic, ginger, turmeric and sliced green chillies for the sort of European-Indian kraut that I love. Mixing red with white cabbage results in kraut that is the most beautiful shade of mauve.

Pack all the cabbage and its briny juice into sterilised jars, ensuring that all the liquid released from the cabbage rises up high enough to cover the kraut. Leave an inch or so at the top of the jar empty for expansion. If the cabbage liquid doesn't rise to the top, you can add a little filtered water – leaving the kraut unsubmerged runs the risk of moulds forming on the surface.

Leave the kraut on your kitchen counter to ferment naturally for a few days. You will see little gas bubbles rising and these can be released by opening the jar each day. When you are happy with the tanginess of your ferment, transfer it to the fridge.

› **Kimchi:** if I had to choose between sauerkraut and kimchi at gunpoint, kimchi would probably just edge it. Napa cabbage leaves are brined for around five hours, then seasoned with chilli flakes, garlic, ginger, dried shrimps, etc. and packed away to ferment. There are literally hundreds of varieties of the Korean ferment, the most common being a cabbage kimchi called 'baechu' and a radish kimchi called 'kakdugi'. I dare not provide a kimchi recipe for fear of it not being authentic, but in essence the process of making it is not dissimilar to sauerkraut.

I particularly enjoy coming home after a long shift at work and rustling up some Kimchi Fried Rice (see p. 222), which is as simple as it is comforting. My recipe is a little different to the authentic

Korean version called kimchi-bokkeum-bap, but it can certainly be adapted, depending on what is lurking in the back of your fridge and the depths of your freezer.

The Koreans love kimchi so much that there is even a museum in Seoul dedicated to the condiment. The average Korean eats 26 kilograms of the stuff annually; to put that into context, it is roughly half the body weight of the average Korean female. The Koreans have even invented specific kimchi refrigerators, with custom storage containers that maintain the perfect temperature and prolong longevity.

Would you believe that kimchi has even travelled to space? South Korea's first female astronaut, Yi So Yeon, entered space with a tin of special kimchi to hand. Amazingly, the Korean space agency allegedly spent a small fortune developing kimchi that was suitable for consumption in outer space, not spoiling on exposure to cosmic rays and not pungent enough to put off other astronauts unaccustomed to its whiff.

› **Kombucha:** a fermented drink made from sweetened tea and a specific culture called a SCOBY, which stands for symbiotic culture of bacteria and yeasts. The bacteria and yeasts convert the sugar into ethanol and acetic acid, which is what gives kombucha a recognisable sour taste. To make kombucha, tea and sugar are steeped in boiling water and cooled before adding the SCOBY. The mixture is left covered for a week to ferment and then poured into an airtight container with more sugar and fermented further for a couple of days, at which point it becomes progressively more fizzy. Fruity, spiced and floral additions can be incorporated for interest, and elderflower seems to be the preferred option for many.

› **Live yoghurt:** I once heard a rumour on the school bus that our school chemistry teacher always made yoghurt at home herself, never opting for the shop-bought stuff. Rumour had it that her airing cupboard, where the hot-water tank was housed, was full of steel pots in which yoghurt was being incubated on a daily basis. To this day, I don't know if there is any truth to this story, but it certainly captured my imagination, and when I got home I told my mother about it. She wasted no time in telling me that in her home in Pakistan, yoghurt was always made from scratch, using large earthenware pots called *kunde*. Apparently, it tasted sublime – rich and tangy in equal measure and quite unlike the monotone shop-bought yoghurt varieties.

Yoghurt production is both economical and deeply satisfying when you know how to do it, and I encourage you to give it a go. You'll kick yourself for all those years of picking up yoghurts on trips to the supermarket.

The process starts by buying a 'live' yoghurt from the shops (it will say 'live' or 'bio live' on the packaging). This is your starter culture, and it needs to be at room temperature before you begin. You need the jars that you will make your yoghurt in to be heated up and ready to go. I often use a wide yoghurt-making Thermos or hotpot that retains the heat. Now heat a litre of full-cream milk to 82°C to kill off the bacteria. Allow the milk to cool to 46°C and then add a tablespoon of the yoghurt starter culture and mix everything thoroughly with clean cutlery before pouring it into your preheated jars.

The most challenging part of making yoghurt is incubating it. It needs to be held at between 43 and 46°C for three to eight hours until it sets. The higher temperature will set more rapidly than the lower temperature. (Thinking back, it makes perfect sense now why an airing cupboard would work so well.) You can also use your microwave as an incubating space with a hot-water bottle or heat pad inside (by wrapping the hot-water bottle or pads around the yoghurt), and lots of modern ovens have functions whereby the temperature can be held at 46°C.

Once your yoghurt is set, transfer it to the fridge. You can use a tablespoon of it to make your next batch of yoghurt if you wish. I adore blending homemade yoghurt with ripe alphonso mangoes, ice and fennel seeds to make a refreshing mango lassi in the summer. I also love mixing it fifty-fifty with soda water and adding dried mint and salt to make a cooling Persian drink called doogh.

› **Kefir:** this is a cultured, fermented milk drink that hails from the mountainous climes that divide Asia and Europe. It is not dissimilar in taste to tart varieties of yoghurt, but is slightly fizzy due to carbon dioxide, the end product of the fermentation process. It is made using kefir grains (small, gelatinous beads containing a variety of bacteria and yeast) submerged in cow's milk and kept at room temperature for a day or so. I tend to buy kefir rather than make it myself at home as there is potential for it to make you ill if made incorrectly. Surprisingly, it takes well to being used in desserts and I have designed a rather wonderful, wobbly panna cotta with it (see p. 224). But you can just as well drink it on its own, pour it over your granola or muesli, mix it in a smoothie, use it to make salad dressings, ice cream, parfait and more.

› **Labneh:** a luscious, thick Middle Eastern strained yoghurt cheese that makes Greek yoghurt look like Cinderella's ugly stepsister. Yoghurt is hung in a cheesecloth for a day in the fridge, after which it turns thick enough to form into small balls, which can then be rolled in herbs and spices and submerged in olive oil. I use it as a substitute for cream cheese or mascarpone, and it makes the most wonderful cheesecake imaginable. But my favourite way of eating it is with tomatoes on toast, as in the recipe on p.221. I sometimes toss it into spaghetti with some za'atar for an effortlessly flavoursome pasta dinner or add chunks of it to salad as a substitute for feta cheese. I even used labneh in the semi-final of *MasterChef*, where I served labneh balls rolled in parsley with lamb shanks cooked in date syrup and aubergine couscous – a winning combination.

- **Natto:** a rather stinky fermented soybean condiment most popular in the eastern parts of Japan. It looks almost like tiny brown jelly beans swimming around in stretchy pale goo (not dissimilar to okra) and is usually eaten for breakfast with rice, chives, karashi mustard and raw eggs. I must admit that this particular probiotic is an acquired taste that I have yet to learn. Legend has it that Samurai Minamoto no Yoshiie was on a campaign in north-eastern Japan between AD1086 and AD1088, when one day, the clan was attacked and the beans that were boiling to feed their horses had to be hurriedly packed away in straw bags. When they were opened a few day later they had fermented to form natto, which appealed to the taste of the soldiers.

- **Tempeh:** a fermented, high-protein plant-based food made with soybeans, this is not dissimilar to tofu, but hails from Indonesia. Soybeans are partially cooked, cooled and then inoculated with fungal cultures that ferment them. As the tempeh ferments, the micro-organisms develop a pale mat of mycelium (the vegetative part of a fungus) around the beans to form a firm cake. It tastes slightly nutty, almost mushroom like, and has to be cooked before eating, which technically means that it is not actually a probiotic, as the cultures are deactivated by heat. But in its defence, tempeh does possess prebiotic fibres which theoretically promote gut health. I tend to cut it into cubes and fry it, before serving with spicy Indonesian sambal or kecap manis (sweet soy sauce). But you can also grind it to grains the size of rice and then use it is a plant-based alternative to minced meat if you prefer.

A word of caution: the word 'probiotic' has become a bit of a marketing buzzword, and many food products that you consider to be probiotic may, in reality, contain no live bacteria at all. Sauerkraut and kimchi in particular can often be totally sterile, especially if bought in vacuum-packed containers or sterilised jars. So, proceed with some caution.

With all of this in mind, it is important to remember that the science of the microbiome and probiotics is still in its infancy. Each year we make great advances in our understanding of this exciting field, but there are still a huge number of unknowns. The benefits of probiotic foods have been vastly overstated on social media and the Internet, given that the majority of clinical research has been conducted on probiotic supplements rather than particular food groups. For instance, a multitude of studies have examined whether probiotic supplements help in IBS, but very few specifically looked at the effect of, say, cheese or kimchi. But you'll still find a thousand people on Instagram touting kimchi as a sure-fire way to cure IBS. Probiotic foods are not some sort of miracle cure or 'superfood', but they are a) yummy, b) highly likely to be harmless and c) pretty likely to do you some good by diversifying the composition of your gut bacteria (although if you already suffer from IBS or any other bowel disease, do consult a dietician or doctor before introducing probiotics into your diet).

I have created some recipes using prebiotic and probiotic foods (see pp. 217–225), which will help you develop a taste for these food groups and introduce you to their benefits. You'll notice that in some of them I tell you to apply heat to the probiotic foods, which, as you may have guessed, is likely to denature much of the bacterial content. This is intentional. I believe that in order to enjoy pre- and probiotics in their raw form, it is always sensible to develop a taste for the ingredients first, and the best way to do that is in their cooked form. These recipes serve simply as a starting point; once you have mastered them, you will have a foundation upon which you can build and experiment independently in the future.

Our ancestors never had fridges

Have you ever considered how much refrigerators have changed the way we eat? We now obsess about food passing its sell-by date, living with the mindset that bacteria on food is synonymous with danger. Huge mountains of food go to waste every day, often with their sterile packaging unopened. Now, I am definitely not telling you to cook up that steak that you forgot to eat two weeks ago, but instead I want to let you know about the world of bacteria out there that are usually completely innocuous and often quite beneficial.

We are taught from childhood that the process of 'decay' is bad, and our over-reliance on hyper-hygienic, mass-produced food has taken away our ability to appreciate the diverse world of fermented foods and good bugs on the food around us.

There is a deep sense of pleasure and gratification in the process of fermentation – a feeling of being truly connected to your culinary creation. You really are cultivating a live organism, and the level of attachment some people feel for their sourdough starter is similar to what some people feel for their own children. Just compare the comments on some of the sourdough forums online with those on Mumsnet and see if you can tell the difference.

I was raised in Manchester until I was five years old, and up to that point my appreciation for fermented foods was, understandably, non-existent. However, when I turned six, we moved to Saudi Arabia, where my parents had found work. We were exposed to Middle Eastern cuisines that I had previously never even heard of, and in culinary terms this was a thrilling time for me. In hot climates like the Middle East, where food can easily spoil if left unattended, fermentation is considered a necessity. I learnt to love fermented foods like pickled beets and turnips, sour yoghurt drinks like doogh, ayran, and creamy labneh, all of which became part of our daily lives. Thinking about their delicious acidic tones, even today, makes me go a bit misty-eyed.

Fermentation is the process by which, historically, we have harnessed the power of good bacteria to our advantage. When you really think about it, there is no food culture on the planet that does not ferment in some way or another. Even Western cultures use fermentation in bread, beer and cheese (to name just a few). But some food cultures take fermentation to the next level. For example, the Sudanese ferment sorghum, pearl millet, dates, honey, milk, wild plants and much, much more – perhaps as a response to a climate that can quickly and without warning bring famine and food shortage.

I had Nutella spread over toasted sourdough bread for breakfast this morning, alongside a cappuccino. It would be quite easy to overlook the fact that all three components – coffee, chocolate and sourdough bread – are made with the help of fermentation. The 'good' bugs that cause fermentation have, for millennia, worked hard to make the foods that humans enjoy, and at some level, our existence depends on them continuing to do this important work. Realising this is paramount in our quest for digestive health and happiness.

Fermentation has become an in-vogue pastime in many culinary circles. During the coronavirus lockdown in 2020, I was delighted to see so many people posting pictures of their fermentation projects and homemade sourdough loaves. I suppose the art of fermentation is experiencing a cultural revival of sorts, which is great because it really is an enjoyable thing to do. There is something grounding about cultivating and caring for a sourdough starter – a sort of rekindling of a long-lost relationship with nature. Perhaps that is why so many of us took to it while in the throes of a global pandemic, when nature was very busy trying to get rid of us.

Another part of the appeal of fermentation lies in the fact that with fairly minimal effort you can create a variety of delicious condiments which then add huge amounts of flavour to your dishes. In walking a tightrope in the grey area between fresh and rotten, there is a sense of both anarchy and control. Each ferment will taste slightly different, even when you have followed the same method. And what works one day might fail the next. The taste profiles are edgy, stimulating, with a sour, lactic note often predominating. Handily, fermentation not only enhances the flavour of ingredients, but also increases their nutritional content and shelf life.

As I described earlier, much of our current interest in fermented foods from a gut-health perspective, comes from their probiotic content. The live microorganisms within them can establish themselves in our gut reservoirs and enhance our metabolic capabilities, strengthen our immunity and bring other benefits to our regulatory physiological functions. If the media is to be believed, it's time to forget apples because 'a ferment a day' is what's going to keep the doctor away.

However, while I remain a huge supporter of including more fermented foods in your diet, there are a few noteworthy caveats to keep in mind. We still don't know all there is to know about the advantages of fermented foods, and because of that, we shouldn't draw any hard conclusions about their health benefits, at least until more research has been conducted. So keep in mind the following:

- Not all fermented foods contain live organisms. For example, beer and wine undergo steps that remove the yeasts which fermented the hops, and many other fermented foods are heat-treated, deactivating the organisms. Bread is baked and sauerkraut is often canned. So, while these foods are nutritious and delicious, they are devoid of microbial content. I am a regular at a local Eastern European delicatessen just a stone's throw from my home, where there are aisles lined with fermented products in jars and tins, alongside a fresh food section where you can buy fresh olives, sheep cheese, live yoghurt and sauerkraut. To get any value from live microbes, I would recommend opting for the fresh food section over the jarred and tinned items.

- The benefits of probiotics are specific to the strain of bacteria that is being assessed. In contrast, fermented foods may contain a number of different bacterial strains, the stability of which is variable from one batch of a product to the next. Some laboratory and animal studies, using fermented foods like sauerkraut or kimchi, have shown encouraging results, but looking at the fine print it seems that researchers often use an extracted, isolated bacterial strain from the huge number of bacteria available in the fermented food.

› It is sometimes tricky to unpick whether the health benefit from a particular fermented food is through the microbes it possesses, or whether another nutrient or chemical is actually the valuable ingredient. Studies looking at cohorts of people show that yoghurt and other fermented dairy foods have been associated with health benefits, but that unfortunately isn't absolute proof of the advantages of the microbes, just proof that *something* in these foods is helpful.

Where do you find yourself on your journey to digestive health and happiness after reading this chapter?

For me, it is deeply humbling to know that we are not alone in our bodies – that living deep within in each of us are trillions of other life forms. Emerging evidence suggests that the vast ecosystem of organisms in our guts may be linked to a plethora of diseases and conditions – which is why learning how to care for our gut bugs is really very important in the long term.

It is undeniable that the food we eat exerts a huge influence on the composition of the gut microbiome. I like to think that through the food I choose to ingest, I am, in fact, communicating directly with my gut bugs, nourishing them, creating a comfortable environment for them, allowing them to thrive; and in turn, they are helping me to flourish. Thus, eating healthily and happily is perhaps as much about feeding our gut bugs as it is about feeding ourselves.

A diverse gut microbiome has become synonymous with good health these days, but there is a long way to go before we can categorically say there is an optimal way of boosting gut microbial health. For now though, there is no harm in finding gastronomic happiness through steadily incorporating more plant-based foods, prebiotics, probiotics (although the science can't tell us definitively how helpful these are just now) and gloriously bug-laden fermented foods into your diet.

Summary

- The trillions of microbes that inhabit our gut are collectively known as our 'microbiota'. The term 'gut microbiome' refers to the collective genome of all the microbiota present in the gut.

- The gut microbiota develops from birth onwards as the foetus passes through the vaginal canal. Breastfeeding is hugely beneficial in maturing the early gut microbiota.

- Progressive research is pointing to a relationship between the composition of the gut microbiota and obesity. This may in part explain the observation that two people can eat the same amount and gain different quantities of weight.

- The microbiome is under the influence of diet. Increasing the variety of plant-based foods, as well as incorporating 'prebiotic' and 'probiotic' foods within the diet may well help cultivate a healthy, diverse microbiome.

- Fermenting is fun, and you can make delicious, multi-dimensional foods, while at the same time rekindling a relationship with nature. There is a sense of controlled disorder in the fermentation process which is deeply soothing for the soul: a sort of restoration of order among the chaos life creates.

- **Not all fermented foods contain live bacteria, and their probiotic potential varies. We are waiting for the science to catch up with the popularity of fermented foods, but although no definitive claims about their health benefits can be made, the current feeling is that they are likely to help foster a healthy, diverse microbiome with all of the attendant advantages to health. In the vast majority of people, they will do no harm, and are likely to be responsible for some good ... so go for it!**

- **We've been gifted all of these 'good' fermentation bugs, so put them to use! You will need the right equipment and a bit of practice to start out. It is trial and error at first, but once you get going, there really is no looking back. True digestive health and happiness awaits ...**

For more information on how fermentation is done properly, refer to the bible of fermentation, *The Art of Fermentation* by Sandor Ellix Katz. It's a must-have for any fermentation enthusiast.

Prebiotic Tabbouleh

Serves 6

Cruciferous vegetables like broccoli, cauliflower and cabbage are prebiotics, i.e. the perfect chow for gut bacteria. They are high in fibre and studies suggest that consuming them will alter the composition of human gut bacterial communities in the large bowel. Sadly, many of us are not great fans of the slightly farty pong they impart. But this tabbouleh recipe which contains raw cauliflower and broccoli changes *everything*! I am sure the 'tabbouleh police' will be out in full force, but hey, don't knock it till you've tried it.

Ingredients

250g broccoli florets
250g cauliflower florets
40g fine bulgur, soaked in boiling water for 30 minutes
125g fresh parsley, washed
30g mint leaves, stalks removed
30g dill
1 medium-sized white onion, finely diced
2 tomatoes, finely diced, seeds and juice retained
1 teaspoon Lebanese allspice powder
Juice of 3 large lemons
1 tablespoon pomegranate molasses
4 tablespoons olive oil
Salt, to taste

Method

1. Grate the broccoli and cauliflower into a large bowl. You can use a food processor, and this is certainly less time consuming, but I feel that grating the vegetables somehow creates a superior texture.

2. Rinse the fine bulgur in cold water till it runs clear, strain through a sieve and place it in another bowl.

3. Now chop the parsley carefully, so as to not bruise it. (You have to slice the leaves, rather than chopping and crushing them under your knife.) A mezzaluna is not desirable here; rather, a

sharp knife will do the trick. Slice the mint and dill finely and add all the prepared herbs to the grated cauliflower and broccoli.

4. Fluff up the bulgur-wheat grains before adding them to the vegetable and herb mix, followed by the onion and tomatoes. Give everything a good stir to combine.
5. To season the salad, add the allspice, lemon juice, pomegranate molasses, olive oil and a generous amount of salt. Toss everything together and taste. Add more salt, spice and lemon juice if you wish.

Note: do not hesitate to use other combinations of herbs, depending on what is available. You can also add other vegetables of your choice – for example, cucumbers (with the seeds discarded and finely diced) or diced courgettes, mangetout, green beans or celery. Even diced apples and toasted seeds would be noteworthy additions.

Apricot, Roquefort and Chicory Salad with Grapes, Candied Walnuts, Chilli and Mint

Serves 4

The sweet, tart flesh of apricots and bitter tones of chicory are a perfect contrast to the deeply savoury crumbling chunks of blue-veined, pungent, probiotic Roquefort. You can just as well substitute any other blue cheese of your choice for the Roquefort. Cashel Blue or Gorgonzola are particularly good and are often 'live' with bacteria that are likely to cultivate the bacterial garden in your gut. Be aware that not all cheese contains live bacteria; processed slabs of orange-coloured cheese or tins of Kraft contain virtually no beneficial gut bugs.

Chicory is rich in a fibre called inulin, a prebiotic that may exert beneficial effects on softening stool for those of us who are constipated.

Ingredients

2 heads of chicory, separated
12 purple grapes, halved lengthways
8 fresh apricots, halved and stoned
2 tablespoons olive oil
1 tablespoon sherry vinegar
½ teaspoon red chilli flakes
50g Roquefort cheese
50g walnuts
1 tablespoon honey
10 mint leaves
Salt, to taste

Method

1. Scatter the chicory and grapes on a large platter. Heat a griddle pan to high, and place the apricot halves on it, flesh-side down. After 2 minutes, check to see if they have char lines on them. Once dark lines are visible, place the apricots on the platter on top of the chicory and grapes.

2. Combine the olive oil, sherry vinegar, salt and chilli flakes in a bowl and whisk gently with a fork to combine. Pour the dressing over the chicory, grapes and griddled apricots. Mix gently with your hands to ensure the dressing touches all parts of the salad. Crumble over the Roquefort cheese.

3. Toast the walnuts in a dry pan over a medium heat for around 2 minutes. Once they are starting to take on some colour, drizzle over the honey. Stir well, ensuring each walnut is coated. The honey will caramelise very quickly, so when it starts to become slightly darker, quickly scatter the now candied walnuts over the dressed salad.

4. To serve, top with roughly torn mint leaves.

Note: you can easily use 2 large ripe peaches if you can't get hold of fresh apricots.

Sauerkraut, Pineapple and Herb Salad

Serves 4–6

I first saw a similar recipe to this in a book on American soul-food cookery. The combination of the fermented probiotic sauerkraut and pineapple seemed rather eclectic, and I felt instantly compelled to try it. And boy oh boy, I do not have any regrets. The contrast of the sour, sweet pineapple and the salty fermented sauerkraut is a match made in culinary heaven. We get through heaps of the stuff in our home, with crispy fried chicken thighs, at barbecues or even just on its own. It really is a must-try. I would encourage you to use fresh rather than tinned pineapple to get the most out of this recipe.

Ingredients

450g fresh pineapple slices
1 Pink Lady apple, cored and diced
200g sauerkraut
2 tablespoons mayonnaise
2 tablespoons crème fraîche
Juice of ½ lemon
½ teaspoon white pepper
Handful of chopped dill
2 sprigs of tarragon (optional)
100g toasted pecans, roughly chopped (optional)

Method

1. Chop the pineapple into small pieces, about the size of your fingertip. Mix the pineapple with the diced apple, sauerkraut, mayonnaise, crème fraîche, lemon juice, white pepper and herbs. Season with salt, but cautiously, as the sauerkraut can be quite salty already. Refrigerate if you are not going to eat it straight away. If you wish, you can top with chopped toasted pecans for extra crunch.

Note: you can easily replace the diced apples with a bunch of seedless grapes, cutting each grape in halves or quarters.

Labneh and Tomatoes on Toast

Serves 2

The probiotic labneh is a popular Middle Eastern soft cheese with a more complex flavour than yoghurt and a distinct creamy tang that is the perfect canvas for an array of toppings, from courgettes to smoked aubergines, olive oil and za'atar or honey and walnuts, for a sweet twist. This dish happens to be a brunch favourite of mine and is particularly pleasing in the height of summer when tomatoes ripen and come to life. It is as if the Italian bruschetta travelled to the Middle East, picking up some rather wonderful flavours along the way.

Ingredients

4 large slices of sourdough bread
4 teaspoons extra virgin olive oil
1 garlic clove
200g labneh
180g cherry tomatoes of various colours, sliced in half lengthways
1 tablespoon pomegranate molasses or good-quality balsamic vinegar
1 tablespoon olive oil
½ teaspoon sumac
1 teaspoon pul biber or other red chilli flakes
1 teaspoon dried or fresh oregano (or za'atar, if you have it)
Handful of finely chopped parsley (optional)
Handful of fresh pomegranate seeds (optional)
Sea salt, to taste

Method

1. Brush the olive oil over the slices of bread and toast on a griddle till charred and crisp. Rub the surface of each slice with the garlic clove, taking care not to squash the bread.

2. Spread the labneh over the four slices of toast and top with the sliced tomatoes. Sprinkle over some flaky sea salt. Drizzle the pomegranate molasses and more olive oil over the tomatoes,

followed by a sprinkling of the sumac, chilli flakes, oregano (or za'atar) and parsley and pomegranate seeds, if using. Serve immediately.

Note: this recipe is blank canvas for you to experiment with. Why not char 2 courgettes in a hot oven until they are completely black, then peel away the skin and use the smoky flesh to top these toasts. Or perhaps use smoked aubergines instead? You can fiddle around with the herbs, too – basil is rather wonderful, as are chives.

Kimchi Fried Rice

Serves 4

Oh, I love kimchi. For those who are unfamiliar with it, it is a staple of Korean cuisine, eaten often as a side dish. It is made of salted and fermented napa cabbage leaves called 'baechu', spring onions, radish, ginger, garlic and seasoned with gochugaru chilli powder.

Korean households often make kimchi in large quantities for their *kimjiang*, or annual kimchi-making event, in preparation for the bitterly cold winters. Lucky for us, it possesses probiotic properties, having been fermented by lactic-acid bacteria, which dominate the process, eradicating other nasty bacterial colonies.

Ingredients

4 heaped teaspoons of Lao Gan Ma crispy chilli oil
4 eggs
2 tablespoons vegetable oil
75g butter
2 garlic cloves, finely sliced
4 spring onions, finely chopped
250g prawns, shelled and deveined
450g pre-cooked white rice
100g frozen peas, defrosted
3 teaspoons dark, rich soy sauce
200g roughly chopped kimchi, plus extra to serve alongside the final dish
Salt, to taste
Vegetable oil, for frying

Method

1. Put a teaspoon of the crispy chilli oil into the bottom of each of 4 breakfast bowls.

2. Fry the eggs in vegetable oil in a non-stick frying pan over a medium heat until the sides are crispy, but the yolks are still runny. Carefully place the fried eggs sunny side down in the bowls, on top of the chilli oil.

3. Now add the butter to your frying pan and crank up the heat to high. Add the garlic and spring onions. When the garlic is turning golden brown at the edges, add the prawns and cook through. Ensure that the heat remans high as you want the prawns to fry off, rather than stew.

4. After 2 minutes, add the rice to the frying pan, along with the frozen peas, soy sauce and chopped kimchi. Splash around 50ml warm water into the pan. The steam will soften the rice and bring all the ingredients together. Give everything a good toss to combine, and within the next minute or so the fried rice will be ready. Taste the rice and add more salt if you feel the dish needs it (the kimchi and soy sauce are already salty, so be cautious).

5. To serve, fill the breakfast bowls with the fried rice. Place a flat plate over the bowls and invert them. Lift away the bowls to reveal the chilli oil and fried eggs with their soft yolks cascading over the fried rice. Eat immediately while warm, with some extra kimchi on the side.

Note: I sometimes like serving with sriracha sauce for extra heat. You can substitute prawns with leftover chicken or if you are vegetarian/vegan you can use tofu and replace the butter with 2 tablespoons of vegetable oil. Add leftover vegetables from your fridge, like green peppers or celery, or add tinned sweetcorn. Sometimes I like some chopped pineapple chunks in my fried rice, but I realise pineapple in savoury dishes can be divisive.

Also note that Lao Gan Ma chilli oil, or 'old Godmother' chilli oil, contains a large quantity of fried, crisp chillies suspended in oil and tastes far milder than other types of Chinese chilli oil. If you can't get hold of it but have other varieties of Chinese chilli oil to hand, use half the quantity specified above.

Kefir and Cardamom Panna Cotta with Candied Almonds and Cherries

Serves 4

I love the lusty wobble of this panna cotta. Kefir has a tart, lactic tang and a slight fizz due to the presence of carbon dioxide, an end product of the fermentation process. It tastes not dissimilar to buttermilk. In this recipe, the gelatine is dissolved in hot double cream, which is then cooled before the kefir is added; this prevents the kefir from reaching a high enough temperature for all the good bugs proliferating within it to denature.

Ingredients

250ml double cream
5 tablespoons honey
½ teaspoon ground cardamom
2.5 sheets of platinum-strength gelatine
250ml kefir
150g ripe cherries, stoned and halved

For the candied almonds:
100g granulated sugar
100g flaked almonds
Vegetable oil, for greasing

Method

1. Heat the double cream and honey in a pan over a low heat, until just at boiling point. Remove from the heat and add the cardamom. Allow the cream to cool slightly.

2. Soften the gelatine by soaking it in some warm water. When it is very soft, squeeze out the moisture and drop it into the warm

cream mixture. The gelatine should dissolve immediately; if it does not, the cream mixture needs to be heated up ever so slightly until it does.

3. Once the gelatine has dissolved in the double cream, allow it to cool for a few minutes before combining the kefir with the cream mixture. Give the mixture a final taste; you can add more cardamom or honey at this point if you wish. Pour the panna cotta mixture carefully into four individual ramekins or dariole moulds. Transfer to the fridge to set for at least 4 hours, ideally overnight.

4. Meanwhile, make the candied almonds. Melt the sugar in a pan over a medium heat until it turns to a deep brown-coloured caramel, swirling the pan occasionally to melt evenly. Quickly add the almonds, stirring once to ensure they are evenly coated, then pour the whole mixture on to a piece of greased baking parchment, spreading it as thinly as possible. Set aside to cool.

5. Roughly chop the candied almonds and sprinkle over the smooth, wobbly panna cotta, alongside the stoned cherries. If you have used a dariole mould, dipping it in boiling water for 10 seconds before turning the panna cotta out will help to detach it easily from the mould.

Note: you can use any berries of your choice, or passion fruit seeds would be particularly delicious.

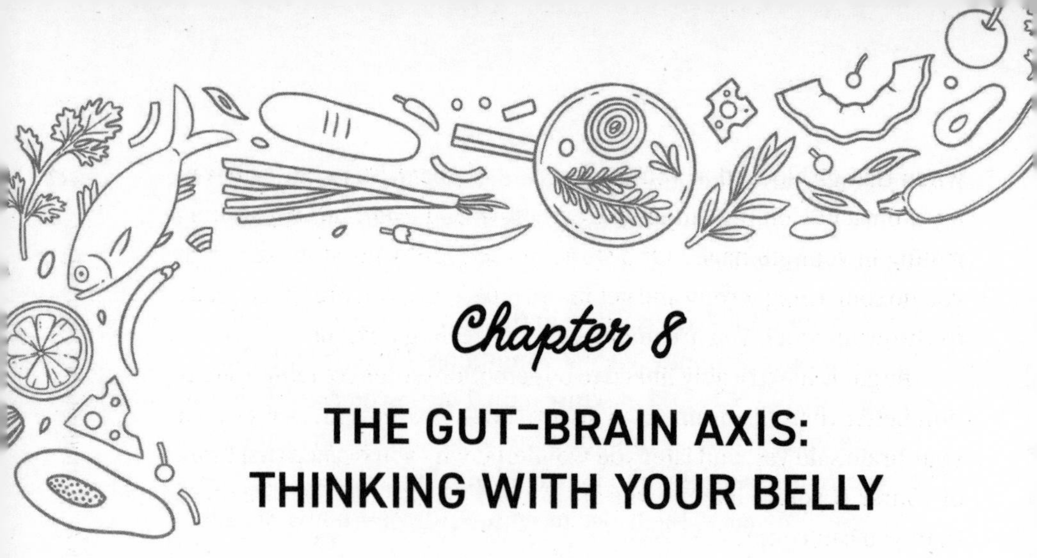

Chapter 8

THE GUT–BRAIN AXIS: THINKING WITH YOUR BELLY

Most people have no idea how much of an impact the gut has on our day-to-day lives, beyond simply digesting our dinner. As we'll see in this chapter, although the brain and gut are located miles away from one another anatomically, there is mounting evidence that they speak to one another, and that this relationship is mediated by our gut microbes.

As discussed previously (see pp. 200–214), the composition of our gut microbes is influenced heavily by the food we choose to eat. And part of finding a sense of digestive health and happiness lies in accepting the fact that what we eat has far-reaching consequences, not just for our digestive systems, but critically, also our mental wellbeing and perhaps even our behaviour. Twenty-first century science is telling us that the saying 'you are what you eat' may, in fact, have more truth behind it than we could ever have imagined.

Gut feeling

We all know the phrases: how did you know your boyfriend was cheating on you? Oh, I had a gut feeling. How did you feel about missing out on Glastonbury tickets this year? Gutted. What was it like when that girl looked at you from across the bar? I had butterflies in my stomach! Coming out of an exam, have you ever had that sinking feeling in your stomach? Or a stone in the pit of your stomach when you do something wrong and get found out? Have you ever been ready to throw up when you are under stress or feeling anxious?

The gut is inextricably linked to our emotions, often reading a situation better than the brain. How many times has your gut said no, but your brain said yes, and later you wondered why you trusted that brain of yours? If you're like me, the answer to that question is more times than you can count.

I recall a time in my university days when a dear friend of mine was seeing someone I did not like at all. They dated for four long years, and a key feature of their relationship was experimenting with every cuisine, eating their way around town, from holes in the wall selling Ethiopian takeout to Michelin-starred restaurants. In particular, the two of them would go to Borough Market in south London every Friday to find something mouth-watering to eat. It was a cute little ritual of theirs, but one Friday, after a particularly heated argument, he took one of his other female friends to the market for lunch. This was the ultimate offence, and soon afterwards, they were both single.

Now, I'd always had a gut feeling about this boyfriend, as had a number of our other friends. This wasn't based on a conscious thought, but just something about him that made us feel uncomfortable – an instinct, felt in the belly rather than the brain.

Information transmission between brain and bowel

As a discipline, medicine conceptualises the body as a group of separate systems functioning independently of one another. We have the nervous system, the digestive system, the cardiovascular system, and we have experts in each of these. But the reality is that many systems of the body are closely interrelated, and a far more holistic approach among the range of currently separate disciplines is now being called for by a number of leading medical experts. In particular, studies are now showing that your brain has a direct effect on your gut health, and that your gut affects your brain's health in ways that we are only just beginning to understand. This bi-directional communication system, between the gut and brain, has been named, rather appropriately, the gut–brain axis.

As you may remember from the previous chapter, a very complex community of microorganisms, our microbiota, covers the inner wall of the human gut (see p. 192). It is a community with which we, over millennia, have formed a very close, symbiotic relationship. Over the last few decades, researchers have discovered that the interaction between the gut and the brain is governed, at least in part, by these microbes, rather than the brain alone. What this means is that our gut microbes affect brain function, and therefore, in turn, our behaviour. Now, you might think it's common sense to say that the gut and brain are somehow linked, but the question is how? And to what extent? And what are the mechanisms behind this relationship?

Broadly speaking, there are three ways in which the gut and the brain communicate with each other: through our nerves, our hormones and our immune systems.

Nerves

It might surprise you to learn that we all possess a second brain in the gut (sort of). Scientists call it the enteric nervous system, and it is made up of around 500 million or so nerve cells, which is not an insignificant number – in fact, it's the same amount as there are in an entire cat brain. So far, around twenty different types of nerve cells have been identified in this enteric nervous system, and they work together to form a pretty sophisticated neural network.

The enteric nervous system in the bowel communicates with the brain via the vagus nerve. This nerve functions a bit like WhatsApp, firing messages in both directions: brain to gut, gut to brain. The trillions of microbes in the gut can stimulate the enteric nervous system, which then sends signals to the brain via the vagus nerve. These signals can affect our emotions, our thoughts, even our actions.

The vagus nerve also engages with the limbic system, an area in the brain which is responsible for reward and the regulation of emotions. The limbic system sends instructions and messages from the brain, which influence the activity of the gut, connecting our bellies to our emotions. This connection, between our emotion centre and our guts, via the vagus nerve, is why we can feel emotions in our bellies, and why it's usually right to trust your gut instinct.

Hormones

The lining of the gut is peppered with a number of cells, called endocrine cells, which specialise in the production of hormones. Your gut is actually, by some estimates, the largest producer of hormones in your body. Gut microbes have been shown by researchers to stimulate these endocrine cells in the gut to produce various substances which reach the brain via the bloodstream, or exert an effect directly on the enteric nervous system. You might have heard of one of these chemicals, serotonin (aka the happy chemical), which is involved in the perception of pain, regulation of sleep, mood, appetite and overall wellbeing. The gut makes nearly 95 per cent of our bodies' serotonin, and even acts as a reservoir, storing it until it is needed.

So, with the gut being so integral to the production of serotonin, it should be a bit clearer now why achieving digestive health and happiness with a gut full of unhappy microbiota is about as improbable as pigs flying. It just won't happen. And bear in mind also that serotonin is responsible for regulating the bowel contractions that help us poo, and it's impossible to not feel calm after a really good poo.

The immune system

The immune cells residing in the gut actually make up the largest part of the immune system, so it's no surprise that many medical professionals refer to the gut as the largest immune organ of the body. But the gut microbiota can affect immune cells in the bowel: stress can tip the microbial balance in the gut, reducing the strength of the immune system and making the host (that's you) more vulnerable to infectious disease. A healthy microbiome, on the other hand, protects the body from harmful bacteria (and subsequent nasty infection), and regulates the immune system, priming it to effectively combat illness.

But what happens if a compromised immune system lets bad bacteria take hold in the gut? If this happens, inflammation of the gut can occur, leading to a loss of its barrier function. This issue, the so-called 'leaky gut' syndrome, is where the gut wall becomes porous and allows bacteria or toxins to enter the bloodstream, causing headaches, confusion, fatigue and difficulty concentrating. In addition, other gut chemicals can escape and bind to the vagus nerve, reducing energy, increasing pain sensation and decreasing feelings of fullness after meals. Interestingly, 'leaky gut' is not yet recognised as a disease, but with more and more evidence emerging, the medical community is beginning to shift its view.

Feeding the gut–brain axis

Just a decade or so ago, the idea that microorganisms in the human gut could influence the brain would have been (and was) dismissed by many as the ramblings of a mad scientist. Today, we stand corrected. There is now ample and ever-expanding knowledge that the microbial community within us plays a vital role in mediating and regulating gut-brain communication. And, when you look at all the evidence confirming the gut's ability to communicate with the brain, it becomes almost unthinkable that this microbe-mediated interaction doesn't play a very important, perhaps critical role in determining an individual's brain function and behaviour.

We also now know that different species of microbes prefer particular foods. Some love sugars and fats, others love fibre or dairy. The depths of your colon are an unforgiving place, where different populations of microbes are doing what they can to ensure survival.

The theory goes that if a particular bug is lacking in its preferred food, let's say it needs fibre to thrive, it can trigger the release of chemicals which, in turn, trigger electrical impulses to travel up your vagus nerve, which can then trigger your brain to choose a salad at M&S instead of a Big Mac. It's also a vicious cycle; if you eat unhealthy food, you run the risk of increasing the ratio of the bacteria that prefer to feast on that unhealthy food in your gut. If you then stop eating that particular food, the bacteria can release chemicals that actually make you crave the food once again, or so the science seems to suggest. When you think about it, this is huge; we are now beginning to realise that our food choices might not be completely our own as the microbiota plays a vital role in regulating different aspects of eating-related behaviour.

Luckily, the composition of the microbiota is not static, and bacteria that dominate one week can be relatively scarce the next. So, while the chicken-nugget dinner we chose to eat a few nights ago can encourage certain less favourable species to thrive and others to wither, if we override cravings for these energy-dense, calorific foods on as many occasions as possible, we can, over time, effectively rebuild our microbiota with a healthier ratio of good to bad bacteria. Whether this knowledge is enough to stop us picking up that second jam doughnut or eleventh chicken nugget, though, is another question entirely.

Depression and the gut–brain axis

The fields of microbiology and psychiatry have not been natural bedfellows, but there have been a few notable exceptions. In 1908, Élie Metchnikoff won the Nobel Prize in Physiology for suggesting that bacteria in fermented milk were beneficial against 'autointoxication', a term which, back then, was used to describe symptoms such as fatigue or melancholy. And Dr Henry Cotton, a New Jersey-based psychiatrist in the early 20th century, was certain that dental bacteria were related to his patients' conditions. He was so certain of this that he would have their teeth pulled out as part of his treatment regimen.

With the development of new technologies capable of mapping human microbes, the old notion that mood conditions such as depression or bipolar disorder start and end in the brain has been increasingly questioned, and the roles of the microbiome and gut–brain axis have been brought ever more into the spotlight. Depression, for instance, is now being considered as a type of chronic, low-level 'inflammatory state'. Those suffering with the condition have been shown to have higher levels of inflammatory molecules in the bloodstream, caused (according to new research) by the barrier function of the gut being compromised, so that microbes within the gut leak into the bloodstream, leading to an inflammatory response.

One limiting factor is that much of the research to date has been carried out on animals, and not humans. And while the evidence is fascinating, we need to wait for human studies to shed further light on the role of microbiota in human depression. What we do know is that rats considered depressed after being separated from their mothers showed a big reduction in symptoms after being treated with the probiotic Bifidobacterium infantis, together with an antidepressant.

Crazily, transplanting poo from depressed rats into others lacking a microbiome led to 'anxiety-like behaviours'. This is staggering, as it suggests that not only is there a link between the microbiota and depression, but also that, at least in animals, the gut microbiota may actually be a major contributing *cause* of depressive states.

As I've said, further human studies are needed to clarify matters and interventions based on the gut microbiome are now under investigation. The University of Basel in Switzerland, for example, is planning a trial of faecal transplants in humans in an attempt to restore the gut microbiome and treat depression. However, I have assisted in the endoscopic administration of faecal transplants, and while I am a firm believer in the science behind the procedure, I can tell you that from where I was standing, managing your microbiota through your diet seems preferable to having someone else's poo piped into your bum.

Which brings me to my next point: can diet prevent, or help to lessen, symptoms of depression?

Mood food

Clinical depression is a very real part of many of our lives. It's so common that over a third of us will suffer with it at some point in our lives, and anyone who has done knows how overwhelming it can be, and how little it helps to have people explain how easy it is to 'cheer up'. So let me reassure everyone reading this that I am absolutely not saying 'if you eat this, you won't feel depressed' or anything even remotely like this, because unfortunately, that's not how depression works. There is no magic bullet, and while it would be lovely if there was a superfood that could alleviate mental suffering by changing our guts' microbe ratios, that isn't yet the case. The best we can hope for currently is this: we know from research that our microbiome affects our mental states to some degree. And we also know from other research that diet can affect the make-up of our microbiomes. So, it seems likely that a food that positively affects our gut might also positively affect our mental state.

Now, with that little disclaimer out of the way, looking at many of the studies together, it appears that a diet high in fruits, vegetables, whole grains, fish, olive oil and low-fat dairy, with a low intake of animal foods, has been associated with a decreased risk of depression. In contrast, a diet full of red and processed meat, refined grains, sweets and high-fat dairy products, with a low intake of fruits and vegetables, has been shown by research to be associated with an increased risk of depression.

This sounds intuitive. We have known about the benefits of the Mediterranean-style diet for many years, and when you look at all the studies collectively, they give the clear impression that this style of eating is associated with around a 30 per cent reduced risk for depression. So, even if we'll never be able to move to the Italian coast and eat bruschetta with a man called Marco who rides a Vespa in a loosely buttoned white linen shirt and chinos, we can at least try to eat more like him.

In 2013, the aptly named SMILES (Supporting the Modification of lifestyle In Lowered Emotional States) project in Australia was the first of its kind to investigate the question, *If I improve my diet, will my mood improve*? It remains a landmark study in the emerging world of nutritional psychiatry. By giving depressed participants either a twelve-week Mediterranean diet or the usual social-support interventions, scientists were able to show that the diet resulted in significantly greater improvements in mental health. In fact, 32 per cent of those on the Mediterranean diet were no longer considered depressed by the end of the experiment. And even more convincingly, the people who changed their diets most dramatically also showed the most improvement in their depressive symptoms.

Two key foods in the war against depression are oily fish and fibre. Omega-3, an essential fatty acid, is abundant in oily fish like tuna, mackerel, salmon and sardines. They seem to exert an effect on the microbiome by helping cultivate the gut bacteria that produce butyrate (the short-chain fatty acid which helps to nourish and nurture the cells lining the colonic wall, reducing inflammation and potential 'leaky gut'). So, include oily fish in your diet, but remember that not all fish are 'oily', and a cod fish-finger sandwich covered in ketchup doesn't count. You can experiment with raw and cooked oily fish as I have done in my recipe for smoked salmon with wasabi lemon dressing and pickled ginger on p. 244. The beauty of fish rich in omega oils is that is remarkably easy to cook, taking literally minutes to prepare – a huge advantage to those with a busy lifestyle. Many oily fish are relatively inexpensive as well, so can be added to your shopping trolley no matter what the budget, which is a huge bonus *The River Cottage Fish Book* described the difference between oily and white fish as that between a long distance runner and an athlete (where the oily fish are the athletes). Oily fish live in the 'pelagic' zones of oceans and lakes, being neither close to the bottom or near the shore.

This pelagic habitat makes these fish agile swimmers with streamlined bodies, capable of sustaining long-distance migrations. They spend their entire lives moving towards their meals and thus need to store energy in the form of fat droplets through their body tissues.

Current guidelines suggest that a healthy, balanced diet should include at least two portions of fish a week, including one of oily fish (i.e. 2 x 140g cooked weight).

Dietary fibre, from sources such as whole grains, nuts, seeds, fruits and vegetables, also seems to exert a positive effect on people's mood. The most likely reason for this is their prebiotic power, in that they feed and nourish the 'good' gut bacteria, again by encouraging butyrate production.

Probiotic foods such as live yoghurt, kefir, sauerkraut and cheese may also exert some beneficial effects on low-mood states, but more evidence is needed to confirm the initial research findings. We know that rodents given probiotics showed positive changes in a number of regions in the brain, and that healthy women given probiotics in the form of fermented milk products have shown altered brain activity when it comes to processing different facial expressions. We can infer, then, that altering the gut microbiota in a person is very likely to have a direct effect on their brain function in an observable way.

Another set of great foods to stimulate your gut's good bacteria are those that are rich in polyphenols (a family of plant chemicals digested by gut bacteria). Polyphenols show diverse structures and over 500 different molecules are known in foods. Cocoa, green tea, olive oil and coffee are all great sources of polyphenols, and while their effects are currently being studied, it looks like they increase healthy gut bacteria growth and may even improve cognition and prevent memory decline in older adults. Whether polyphenols have an effect on depressive states remains to be seen, but the information that we currently have looks very promising indeed. Interestingly, many spices and herbs are rich in polyphenols, particularly star anise, cloves, dried mint, dried oregano, cumin and dried ginger, some of which take very well to being brewed for tea, simmered in aromatic Asian-style broths or, as I have used them, cheekily in a sticky glaze for melting, fatty short ribs (see recipe, p. 246).

Lastly, I just want to mention a chemical called tryptophan. It's an amino acid that is used to make serotonin, the 'happy' chemical (see p. 229). Now, the body can't make tryptophan by itself, so we are completely reliant on diet to get enough of it. And given that serotonin is a huge player in the regulation of mood, and you need enough tryptophan in the brain to make the serotonin, it stands to reason that a diet rich in tryptophan-heavy ingredients might increase our serotonin production. Small studies have indeed shown that high levels of tryptophan in the diet may result in fewer depressive symptoms, or less severe ones.

If you want to give your body a boost of tryptophan, tofu (and other soy-based foods), chicken, turkey, egg white, mozzarella cheese, peanuts, pumpkin, sesame seeds and milk are all rich sources of this very useful chemical. I adore using tofu in my cooking, particularly silken varieties where the whey has not been pressed out, resulting in a velvety texture called *kinugoshi* in Japanese. You can use the very soft varieties of silken tofu to make vegan desserts, like mousses, or creamy sauces for pasta (useful if you are following a plant-based diet and want an alternative to cream). The slightly firmer silken tofu varieties can be cut into thick cubes and then submerged in soy sauce, sesame oil, chilli and spring onions to make a beautiful five-minute starter or can be seasoned and then baked in the oven as I have done as part of a magnificent tryptophan-rich noodle recipe on p.249.

Science aside, I am a firm believer in the therapeutic benefits derived from the very act of going into the kitchen to engage in an activity that is fundamentally creative. So many times, I have returned home after an awful hospital shift to find solace and comfort in the joyous process of rustling together a fresh meal. Eating the right foods may very well give our gut the edge it needs to keep our serotonin levels up, but actually cooking them gives me purpose, grounds me when I feel things getting on top of me and lifts my mood immeasurably. It's an experience that I am certain will resonate with many readers, and it's why I can tell you, from my own experience, that the best place to find your digestive health and happiness is in the pans, cupboards, drawers and work surfaces of your own kitchen.

Chocolate

I simply could not write about foods that influence our mood, without mentioning our good friend chocolate. Chocolate is with us from the very beginning – we grow up begging our parents for Coco Pops for breakfast, and we go to school reading about Charlie and his chocolate factory. Then, in adulthood, we choose our favourite flavours of Quality Street, Roses and Celebrations, and then judge others by the variety they consider their favourites (or maybe that last bit is just me?). Chocolate is a universal language, too. No matter which country of the world I am in, it brings the same joy to the faces of children and adults alike.

There is no denying chocolate possesses a social value as consuming it is very much a social activity. Chocolate is the language of love, inspiration for countless movies and books. And in some ways, a box of chocolates is symbolic of life itself, given the fact that 'you never know which one you are going to get'. Most recently, I have found myself particularly drawn to the chocolate counter at Fortnum & Mason in London. The bars in the 'Chocolate Library' have been renamed: 'Bittersweet Romance', 'Goodnight, My Bittersweet Beloved', 'The

Forest That Floated', 'Pilots Fly in Pink Skies' and so on. These names may sound somewhat peculiar at first, but each chocolate bar is accompanied by a short story. The identity of the author of these stories is kept a coveted secret, but you can read them on the journal section of the F&M website. For me, they illustrate the immense creativity that chocolate imparts and how in its different flavours and formats, it is able to connect with a multitude of human emotions.

You probably won't need to dig deep before remembering the last utterly fantastic chocolate experience you had, whether it was the squidgy centre of an artisan brownie or the most slippery smooth hot chocolate on a cold winter's day. I remember my husband and I attending a medical conference in Berlin and naughtily slipping out of afternoon lectures early to visit the world-famous chocolate café Rausch Schokoladenhaus. When a boss of mine later asked me what Berlin was like, I started recounting my amazing chocolate experience. After around five minutes of my chocolate monologue, he said, 'Well, I actually meant what were the lectures on anticoagulation agents like? But I am glad you engaged in all forms of education available to you.' Blush.

Chocolate is prepared, via fermentation, from the fruit of the Theobroma cacao, a tropical tree whose name literally means 'food of the gods' in Greek. Native to the Peruvian city of Iquitos on the banks of the Amazon river, the tree bears large fruits, each around 30 centimetres in length. These are harvested in mind-boggling quantities to meet the immense global demand for chocolate. Growing cacao is an extremely labour-intensive process.

Of all the foods that we eat regularly, chocolate is the richest source of a compound called flavonoids, part of the polyphenol family. As we know, polyphenols are food for our gut microbes, and the more of them our gut bacteria have to munch on, the more they will benefit our physical and mental health.

Any study looking at chocolate is something that we should all pay attention to, and one small study looking at the effect of cocoa polyphenols showed that they increased two very beneficial bacteria in the gut, and reduced the proportion of Firmicutes (the bacteria felt to be associated with increased obesity – see p. 198). Another study compared ten people who ate (dark) chocolate a lot with ten who didn't eat it at all (I didn't even know such people existed). They were given the same diet otherwise, and the research found that at the end of the trial, the two groups possessed different gut microbes, with the chocolate eaters showing larger amounts of 'good' bacteria and reductions in those associated with negative effects. Sadly, the small sample size of both these studies does make it difficult to draw definitive conclusions about the role of dark chocolate on human health, but the results seem promising.

In another small study, eighteen participants were divided into two groups and given a series of word and number tests to complete. Then one group was given 24g of dark chocolate a day (70 per cent cacao), while the other ate around the same amount of cacao-free white chocolate. At the end of the intervention, those eating dark chocolate showed some enhanced brain function when repeating the same word and number tests, whereas those eating white chocolate showed no difference whatsoever. And although the study is *far* from perfect in design, it gives us a suggestion that dark chocolate may confer beneficial effects on alertness and improve brain function. I, for one, would like to see more large-scale chocolate-based studies emerging – just the type of research that tickles my taste buds.

But the real magic of chocolate lies in its chemical complexity. Alongside its benefits for our gut, it also contains a range of chemicals that have been proven to be beneficial to other parts of our bodies. For instance, one of these chemicals is called theobromine, and it increases heart rate and decreases blood pressure. Another is phenylethylamine which, taken on its own, is a psychoactive drug and the starting point for the manufacture of antidepressants. A fruit and nut bar is probably not quite sufficient to exert that sort of effect on the brain, but it's just another example of the nice things that chocolate has inside it.

Chocolate is one of those foods that is almost synonymous with the phrase *guilty pleasure*, with particular emphasis on the word guilty. And because we often associate chocolate with guilt, we can lose sight of the fact that when eaten in moderation, chocolate has the potential to be one of those foods that has a measurable positive impact on gut health, and therefore a measurable positive impact on mood. It is also a snack that people might turn to when they are sad or depressed, as a comfort. The debate as to whether depression causes people to eat chocolate or if people use chocolate to relieve depression continues to rage and will do so for a while yet.

It is a difficult question for scientists to answer in many ways, and I have thought long and hard about the best advice to offer on the matter. If you are looking to find digestive health and happiness through chocolate, sadly, I can't tell you to eat more of it on scientific grounds, as the full extent of the relationship between chocolate and human health is not clear. But, what I can tell you is that in moderation, a square or two of *dark* chocolate a day might do your gut some good, and at the very least it will probably do it no harm.

I do have to stress dark, not milk chocolate in this case. Because although I have stated my love for the occasional Cadbury's Dairy Milk, the addictive 50:50 mix of fat and sugar in milk chocolate does mean it is not quite so good for us as the more bitter, adult dark chocolate varieties that exist. From personal experience I can tell you that with time and practice eating various varieties of dark chocolate, you will find that they really do grow on you, and they possess a far superior, complex flavour profile. Like the late Roald Dahl, my current favourite variety of dark chocolate is Prestat. The 'Dark and Stormy' 73 per cent chocolate bar, made from beans cultivated in Côte d'Ivoire, is beautifully aromatic and an excellent evening companion to my television, sofa and blanket after 9 p.m.

Dark Chocolate, Tahini and Pretzel Pie

Serves 8

This is a recipe that converts those who find themselves on the fence about dark chocolate to veritable dark-chocolate fanatics. The contrast of salty pretzels against dense, earthy tahini and bitter dark chocolate creates a showpiece dessert that is rather difficult to forget. And while I do not advocate eating this every day, there is no doubt that a good chocolate dessert does something for ones soul.

Ingredients

For the base:

150g pretzels

50g granulated white sugar

125g melted butter

For the tahini layer:

2 heaped tablespoons tahini

225g white chocolate

300ml double cream

For the dark chocolate layer:

200g dark chocolate

300ml double cream

120g toasted hazelnuts, roughly chopped

1 teaspoon cocoa powder

Method

1. Line the base of a 23cm loose-bottomed or springform tin with greaseproof paper.

2. Blitz the pretzels using a food processor or by placing them in a plastic food bag and crushing to crumbs using a rolling pin. Mix in the sugar, then transfer to a bowl and pour over the melted butter. Mix thoroughly until the crumbs are completely coated and tip them into the prepared tin, pressing down firmly to create an even layer. Chill in the fridge for at least 30 minutes.

3. For the tahini layer, place the tahini, white chocolate and double cream in a bowl and place the bowl over a pan of simmering

water. Stir gently until the chocolate has all melted and the mixture looks smooth. Pour the mixture on to the pretzel base and transfer to the fridge. Allow the tahini layer to set in the fridge for around 4–6 hours or so.

4. For the dark chocolate layer, place the chocolate and double cream in a bowl and place the bowl over a pan of simmering water. Stir gently until all the chocolate has melted and the mixture looks smooth and glossy.

5. Remove the tin from the fridge and scatter the hazelnuts over the set tahini layer. Now pour over the dark chocolate mixture, smoothing the top with the back of a dessertspoon or spatula. Leave to set in the fridge for at least another 4–6 hours, but ideally overnight.

6. To serve, bring the pie out of the fridge and dust with the cocoa powder. Remove the springform tin by opening the outer hook of the tin and slipping the pie gently on to a serving plate. Remove the base of the tin and lining paper gently. Allow to rest for 15 minutes at room temperature before serving.

So where does all this new-found knowledge about the 'gut–brain axis' leave us in terms of finding digestive health and happiness?

The food we eat has a direct impact on the composition of our gut bugs, and these can, in turn, influence brain function in myriad ways. You can argue that even without scientific proof, it is common knowledge that what we eat can influence our health and wellbeing: after all an entire 'health food' industry has burgeoned off the back of this fact.

But for me, knowing that the science is rapidly catching up and actually proving the mechanisms by which certain foods can impact brain function, behaviour and mood is highly empowering. To find true digestive health and happiness is to know that the food you choose to eat has far-reaching effects. Yes, it all starts in the gut, but the gut talks intimately with your brain and you have the power to influence that conversation through what you choose to eat. Wow!

Summary

- The gut–brain axis is a bi-directional communication system between the bowel and the brain.
- Broadly, there are three ways that the gut and brain communicate with one another: through our nerves, hormones and the immune system.
- The composition of our gut bugs, otherwise known as the microbiota, can mediate and regulate communication between the brain and the gut.
- There is evidence that a Mediterranean-style diet can help with depression. High fibre, oily fish rich in Omega-3, probiotics, polyphenol- and tryptophan-rich foods might all have a role to play in improving mood.
- The act of cooking is in itself uplifting for many of us.
- An occasional piece of dark chocolate will do your mood no harm.

Smoked Salmon, Wasabi Lemon Dressing and Pickled Ginger

Serves 4

There are three different types of Omega-3 fats to be found in food: ALA, EPA and DHA. ALA (or alpha linolenic acid) cannot be made by the body and is found in vegetable oils, nuts and seeds (for example, in walnuts, pecans, hazelnuts and linseeds). The other two omega fats, known as EPA (eicosapentaenoic acid) and DHA (docosahexaenoic acid) are found in oily fish like sardines, salmon, mackerel, herring, oysters and anchovies. They can also be made in small quantities from ALA by our bodies.

Any of the above fish can be substituted here, either very fresh and raw, or grilled and charred to perfection.

Ingredients

250g smoked salmon slices
½–1 teaspoon wasabi paste (use horseradish if you don't have wasabi)
1 tablespoon mayonnaise
1 tablespoon soured cream, crème fraîche or Greek yoghurt
2 tablespoons lemon juice
100g cooked edamame beans
75g samphire (optional), blanched in boiling water for 1 minute, then cooled
1 spring onion, finely sliced lengthways
1 teaspoon rapeseed oil
1 teaspoon sesame seeds
25g pickled sushi ginger

Method

1. Spread the smoked salmon in a loosely folded fashion all over a large flat platter.

2. Mix the wasabi paste with the mayonnaise, soured cream and lemon juice. Drizzle this over the salmon. (Add more wasabi if you wish; the idea is to get a bit of heat at the back of your nose and throat with it.)

3. Mix the edamame beans, samphire and spring onion in a separate bowl with the rapeseed oil. Scatter this vegetable mixture over the dressed salmon and sprinkle over a final flourish of sesame seeds and the pickled sushi ginger. Serve at room temperature.

Masala Spiced Sardines and Lemony Kachumber Salad

Serves 2 generously

The joy of oily sardines is not just in their flavour, but also in how inexpensive they are: yes, I am a sucker for a good bargain! While I don't have any qualms about eating the tinned variety, there is something marvellous about this spiced fresh sardine dish. Sprats, small herring and mackerel would also work well here, if you cannot get hold of fresh sardines from your fishmonger.

Ingredients

2 garlic cloves
1 thumb-sized piece of ginger
1 teaspoon hot paprika powder
1 teaspoon red chilli powder
1 teaspoon cumin seeds
1 teaspoon garam masala
1 teaspoon curry powder
1 tablespoon Greek yoghurt
Juice of ½ lemon
500g fresh sardines (gutted and cleaned)
200g diced tomatoes (retain the seeds)
½ medium-sized red onion, finely diced
Juice 1 of lemon
Vegetable oil, for frying
3 tablespoons plain flour
10 mint leaves, finely sliced
Salt to taste

To serve:

Flatbreads of your choice and green chutney (see p. 158)

Method

1. Crush the garlic and ginger to a paste in a mortar and pestle. Add the paprika, chilli powder, cumin seeds, garam masala, curry powder, Greek yoghurt, juice from the half lemon and salt to the crushed ginger and garlic and mix well. Pour this spicy paste over the sardines and allow to marinate for around 15 minutes.

2. Meanwhile, mix the tomatoes, red onion and juice of the whole lemon in a bowl and season generously with salt. Leave this to rest for 15 minutes, so the flavours develop and the onions loose some of their acridness.

3. Heat a few tablespoons of vegetable oil in a non-stick frying pan over a medium heat. Place the flour in a shallow dish and coat the marinated sardines in it, dusting off any excess before gently placing them in the oil. Do not wipe away any of the marinade before coating with flour – you want to keep the flavour on the sardines. Fry the sardines in batches of three or four, turning once. They are ready when they have a deep golden crust (about 3 minutes of frying on each side).

4. To serve, heat up some flatbreads and top with the sardines and kachumber salad. Serve alongside a punchy green chutney or another pickle/chutney of your choice (e.g. mango).

Sticky Polyphenol Glazed Short Ribs

Serves 4–6

Many people don't realise that spices like star anise, cloves, cinnamon, cumin and ginger are rich in polyphenols, as are fruits, like grapes, apples, pears, cherries and other berries and beverages like red wine, tea, coffee and cocoa. A plethora of research points to them being protective against the development of many chronic diseases, owing to 'antioxidant' properties, as well as having beneficial effects for gut bacteria.

These ribs are a prized cut for slow-cooking and are smothered in a very dark, sticky glaze made with polyphenol-rich spices, particularly cloves. Abundant, naughty and melting, pair them with a lightly dressed crunch slaw and a side of fat chips for maximum pleasure. I would recommend eating them with your hands, rather than a knife and fork. And, if it makes you feel better to know that the fatty, meaty morsels of goodness are glazed in polyphenol-rich spices, well, I suppose that's just a little bonus! If I am being honest, polyphenolic properties aside, I think they taste just divine.

Ingredients

2kg beef short ribs, cut crosswise into 5cm-thick chunks
40 cloves
10 star anise
2 tablespoons ground cumin
2 tablespoons fennel seeds
2 thumb-sized pieces of ginger, skin on and cut into rough chunks
2 cinnamon sticks

For the glaze:
100g tamarind, from a block
6 tablespoons ketchup
6 tablespoons soft dark brown sugar
1 tablespoon cocoa powder
6 tablespoons soy sauce

Method

1. Start by placing the ribs in a very large pot and covering with around 3 litres of water. Add the cloves, star anise, cumin, fennel seeds, ginger and cinnamon sticks and bring to the boil. Cover the pot with a lid and allow the ribs to boil over a medium heat for around 2 hours. You are looking for the ribs to be soft and cooked through, but not quite falling off the bone at this stage. Check them every 20 minutes or so to ensure that the liquid has not dried out (top up with water if necessary). Take off the heat,

and once the ribs have cooled, remove them from the cooking liquid with a slotted spoon and place them in a single layer on a non-stick baking tray.

2. Now to preparing the glaze. You will have some liquid left in the pot that you boiled the ribs in. The amount will vary according to how the ribs were cooked, the dimensions of your pan, etc. Add all the ingredients for the glaze into the remaining liquid and stir everything well. If the liquid in your pan has dried out completely, you may need to add some water (around 300ml or so) back to the pot to get things going.

3. Allow the contents of the pot to simmer down to form a shiny, bubbling sauce with the consistency of double cream. How long this takes will invariably depend on the amount of liquid you have in the pot to start with. Pass this sauce through a sieve and then pour it all over your ribs, ensuring they are coated all over.

4. Place the ribs under a hot grill and watch them caramelise. Be careful to not let them burn. Turn the short ribs every 2 minutes or so and continue to baste them with the residual glaze at the bottom of the baking tray until you are happy with their jammy consistency. There may be some beef fat separating away from the glazed ribs and pooling on your baking tray; this is to be expected if the ribs were very fatty to begin with. It takes around 4–8 minutes or so in the grill until the ribs are fully sticky and coated with the deep dark glaze. Remove the ribs from the grill and serve immediately.

Nom Nom Noodle Salad

Serves 2

A friend of mine describes this salad as 'Nom nom nom', which, I believe, loosely translates to delicious! It is spicy to the point that your mouth feels the heat, but not spicy enough to blow your brains out. I use hoisin instead of plum sauce, as it is that bit more intense, but you could certainly use leftover plum sauce as an alternative.

Chicken, tofu and cashews as utilised in this recipe, are fantastic sources of the essential amino acid tryptophan, the precursor for the 'happy' chemical serotonin. And while I can't guarantee that this dish will cure the blues, it does taste wonderfully uplifting.

Ingredients

350g firm silken tofu
1 teaspoon Chinese five spice powder
2 tablespoons vegetable oil
2 nests of egg noodles
500g shredded cooked chicken
1 Turkish cucumber (or ½ a normal cucumber)
2 spring onions, finely chopped
Handful of finely chopped coriander
1 tablespoon chilli oil
3 tablespoons hoisin sauce
100g toasted cashews

Method

1. Preheat the oven to 200°C fan.
2. Remove the tofu carefully from its packaging, so you have a block of it in front of you. Slice it into 1.5cm-thick slices and lay them on a greased, non-stick baking tray. Sprinkle over the five spice powder and drizzle with the vegetable oil. Transfer to the oven for 10 minutes to cook through. Remove from the oven and set aside to cool slightly.
3. Meanwhile, prepare the egg noodles according to the manufacturer's instructions: they usually need to be boiled for

5 minutes or so, then drained. Scatter the prepared noodles on to a large platter and top with the shredded chicken.

4. Slice the cucumber as finely as you can on the diagonal to create oval slices. Scatter the slices over the noodles and chicken and top with the spring onions and coriander.

5. In a separate bowl mix together, the chilli oil and hoisin sauce to make the salad dressing. Drizzle this all over the noodle, chicken and vegetables and toss everything together to coat evenly. Hoisin is quite salty, so you shouldn't need extra salt.

6. Complete the dish by topping with the cooled tofu pieces and a handful of the toasted cashews, roughly chopped.

   Nom nom nom indeed.

Note: For a vegan alternative, substitute the chicken for stir-fried mushrooms or double the quantity of tofu.

Chapter 9

LOSING THE BLOAT

Until now, we have focused mainly on explaining what happens when your body interacts with the food you eat. We've covered the history of cooking, taste and texture (focusing on a few particularly interesting tastes along the way), why we feel hungry and full, as well as getting to know how our gut microbiome can affect our lives in fascinating ways that we wouldn't have previously expected. For me, these are all extremely valuable in laying the foundations upon which we can build our sense of digestive health and happiness.

In the final two chapters, however, we will concentrate not on explaining what happens when the body works as it should, but rather on what happens when it doesn't. Understanding how and why things go wrong is, to me, equally important foundational knowledge in our quest for a deeper relationship with our food, because it can allow us to confidently recognise when our bodies are telling us that things inside aren't quite right. Plus, it allows me to offer those of you who are suffering from gut-related ailments (and also those who aren't) a few recipes that might help you achieve greater digestive comfort.

Farts are a universal language. There are people who are fluent and confident, and others who would prefer the world to think that they don't speak this most divisive of languages. Studies have shown that babies as young as fifteen weeks will laugh at their own farts (or those of people near by). And it's easy to see why. It relaxes the body, relieves pain and sounds hilarious. But, if you are anything like my good friend Delia, gas can also be debilitating – an uncomfortable secret that you would prefer nobody knew.

Delia and I have been friends since we began to walk, and as with friends of that calibre there is practically nothing that I don't know about her. We had our first periods at the same time, our awkward teenage bodies changed in tandem, even our hairstyles fluctuated between passable and disastrous in step with each other's. Delia was, and is, an extension of me.

After a gruelling set of A levels, Delia and I decided that we had earned a treat. So, seeking a change from a trip to the cinema, we booked ourselves in for a full-body Thai massage. Now, if you've ever been for a Thai massage (and despite what you are about to read, I highly recommend it), you will know that the masseuses are trained to get rid of every ounce of stress from your body. Things started off well. Wrapped in soft, freshly laundered dressing gowns, with towelling slippers on our feet, we felt the stress of exams dissipating.

As we both lay covered in towels, the two masseuses assigned to us started their work. I could feel myself drifting into a deep, serene sleep when, out of the corner of my eye, I caught a glimpse of Delia. Far from looking relaxed, her face bore an expression that I did not recognise: a sort of anguish, creasing her nose, a bead of sweat forming on her brow. Every so often, I could hear tiny, low-pitched groans of discomfort. She didn't seem to be in pain, exactly, but something was clearly off.

As the back massage ended, we were instructed to turn around and face the ceiling. I could see her unease, but decided I would interrogate her later; for now, she assured me that a change of position was all that was needed. The masseuses began working on our stomachs. It's a bit strange having your belly massaged, but it was actually very pleasant. We had warm oil gently rubbed on to our stomachs, by hands moving in a rhythmic clockwise motion.

'Oh no,' I heard from the bed beside me.

I looked over at Delia (usually so cool under pressure that she once forgot about an important presentation until her name was called, yet delivered it off the cuff without breaking a sweat – and got the highest mark in class). She was now clearly in distress. The masseuse, meanwhile, focused on her art and apparently oblivious to Delia's discomfort, kneaded her belly with gentle but increasing intensity, until Delia, unable to take any more, shifted her weight slightly.

It wasn't much of a movement, but it was enough. A huge fart ripped through the tranquil air of the spa. We all looked at each other, not sure how to proceed. Then, like thunder following lightning, an eggy smell joined us at the massage tables, cutting through the calming scent of essential oils. Delia looked at me in complete horror, the bead of sweat on her forehead now dripping down her cheek and off her face. She was frozen, unsure what her next move should be.

However, her masseuse, clearly a consummate professional, seemed totally unfazed by the incident. In fact, she began to rub Delia's belly harder and harder, and with each circular motion of hand on belly another gust of wind escaped. Not content with the results she was getting, in one final move the masseuse raised Delia's legs and pressed her own knee firmly into her abdomen, folding and deflating Delia like an air mattress being put back in the loft, each movement giving rise to yet another gush of bubbling wind.

Once satisfied that her work was complete, the masseuse left the room, leaving Delia and I to process what had just happened. It transpired that since exam revision had started, she had found herself becoming progressively more bloated with stomach cramps that only eased when she was able to pass wind or poo. Her tummy felt so ballooned at times that she had resorted to wearing trousers with elasticated waistbands.

Delia's mother was understandably concerned – less about the embarrassment and more about how it might affect her exam performance. She changed Delia's diet to include either broccoli, cauliflower or beans each night with her dinner. Avoiding dairy and bread didn't really help, but for some reason mint tea and her grandmother's Barbie pink Pepto-Bismol did. Eventually, Delia was diagnosed by her GP with irritable bowel syndrome, or IBS – a condition that can, if not managed effectively, wreak havoc in the lives of those suffering from it.

Stress is a huge factor in the flaring up of IBS, and wouldn't you know it, once her exams were behind her, Delia's bowels returned to normal quite quickly. A few shopping trips, a couple of movie nights and some good meals with family and friends began the healing process, and she even met a dietician who, thankfully, advised her to stop eating broccoli and beans every evening. Her belly flattened, and the tracksuits with their elasticated waistbands were pushed to the back of the cupboard.

Delia developed the same bloating symptoms on a few other stress-filled occasions: when she was planning her wedding, when her dog died and when she was dealing with her own colicky first-born. But managing overwhelming bloating can be a challenge, as any IBS sufferer will tell you, and people tend to find their own ways of dealing with the symptoms. Delia, for instance, found that the Downward Facing Dog yoga position was very effective at quickly releasing her trapped wind. (She found that out in a crowded yoga studio, but hey, at least she found out.)

Delia would have probably benefitted from this chapter, had she been able to read it when her bloating issues first manifested, way back in school. When I showed it to her recently, she told me that something that makes IBS more stressful (therefore leading to more symptoms) is that the topic of bowel gas and other abdominal symptoms is shrouded in shame and embarrassment, and now, when I think about her journey, I realise why it is so important for us, as a society, to understand the concept of bloating and why it exists.

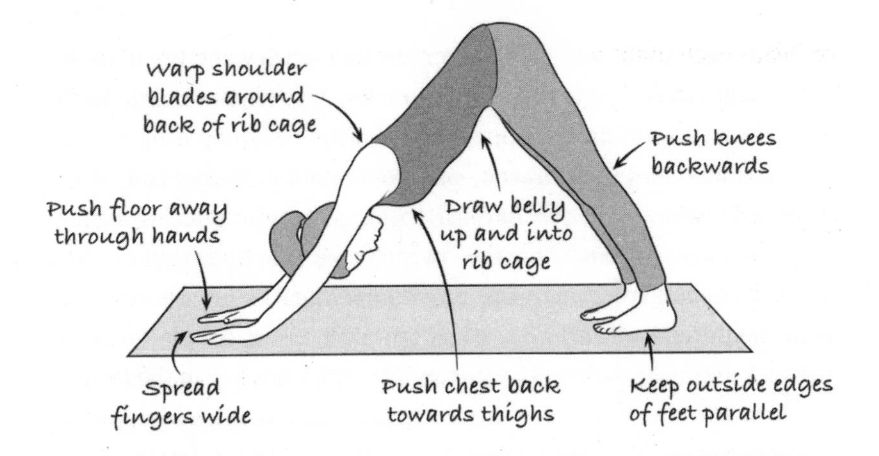

Downward Facing Dog, known in the yoga community as Adho Mukha Svanasana, *is a fantastic yoga move which helps some people reduce feelings of bloating.*

Fascinating facts about flatulence

The human colon is essentially a tiny manure pit, and, as such, bacteria living in it ferment our waste, creating a complex mix of gaseous by-products as part of the process of breaking down our food. Now, most of our food is broken down and digested in the small intestine, but some things, like sorbitol (found in artificial sweeteners), fructose (found in fruits), fibre-rich foods like legumes, as well as complex carbohydrates (like those found in wheat grains) are broken down in the large intestine. It's the gas that is released during the process of breaking down these foods that can erupt from our bums.

But, as ever, the story is not that simple. Wind within the gastrointestinal tract is generated from a range of different sources, not just the whole-grain sandwich we had for lunch. We swallow air all the time, and our bodies are always engaged in chemical processes that create a variety of gases. Estimates suggest that overall, we make around 500 to 1,800 millilitres of bowel gas each day, farting between fourteen and twenty-three times daily. This is almost half of what it takes to inflate an average balloon, and if you enjoy a very high-fibre diet, you might even be able to fill the balloon up all the way.

Farts are as unique to us as our fingerprints. Nitrogen, oxygen, carbon dioxide, hydrogen and methane, all of which are odourless, account for more than 99 per cent of each toot. And less than 1 per cent is comprised of sulphur-containing compounds which are malodourous; if you've ever been to a natural sulphur spring (or cracked open a rotten egg), you will be all too familiar with the smell in question.

There are three key sulphur-containing compounds in our farts: hydrogen sulphide, methanethiol and dimethyl sulphide. Hydrogen sulphide is described classically as having the smell of rotten eggs, methanethiol smells more like decomposing vegetables, while dimethyl sulphide has a sweeter, more sickly tone. The human sense of smell is quite reliably able to distinguish between these three sulphurous stenches, and often individuals will have their own recognisable ratio of smells, their personal 'brand', which is probably why your partner can always tell whether it's you who farted, or whether it really was the dog.

Sulphur-producing bacteria are present in the large bowel in constantly varying quantities, but the most troubling combination occurs when they colonise the right side of the large bowel nearest the rectum. When this happens, those sitting nearest the sufferer will experience a noxious, warm odour, due to the large amount of composting that takes place right near the exit of the bowel. How lovely.

Interestingly, while you may feel overpowered by the smell of a fart that is particularly high in sulphur, it is important to remember that the absolute smelliest of human farts only contain around 1–3 parts per million (ppm) of hydrogen sulphide. It would take 150ppm to paralyse your sense of smell, and over 1,000ppm to cause respiratory paralysis and suffocation. Thankfully, even though it definitely seems otherwise when you are sat next to one, even the most active of bums is unable to produce enough sulphurous gas to permanently damage anyone's health. This is small solace, however, when your partner farts in bed next to you.

If you've ever thought that your farts change their aroma after a steak dinner, compared to after a breakfast of cereal and fruit, you'll be pleased to know that you'd be right. The offensive smell of our gas is a result not only of the composition of the colonic microbiota, but also the food we put through our guts to feed those bugs. And, from studies looking at the social dimensions of farting, we know that it isn't the act of farting publicly that bothers us as much as the lingering smell generated – so I assume that many of you might want to know a bit more about which foods lessen the production of sulphurous odours. For a friend, obviously.

Unsurprisingly, very few researchers want to spend all day smelling other peoples' farts, and because of this, there is a lack of data on which food groups might reduce our sulphurous emissions. The small studies that exist suggest that if you take stool from healthy volunteers and mix it with cysteine, a compound that is found in high quantities in meat, eggs and dairy, then the hydrogen-sulphide emissions from gut bacteria can increase more than seven-fold. This might explain why body builders (who take protein powder supplements rich in cysteine) or people with diets high in red meat are famous for their particularly smelly brand of wind.

However, scientists also noticed that when sulphurous-smelling excrement was mixed with four different 'slowly absorbed carbohydrates', these carbohydrates were preferentially fermented by gut bacteria ahead of proteins, resulting in up to 75 per cent less hydrogen-sulphide production. These slowly absorbed carbohydrates, known as resistant starches, are found in potatoes, bananas, legumes and cereals, as well as in fructans (compounds found in wheat, artichokes and asparagus).If you are worried about the smell of your farts, trying these groups of foods will do you no harm, and you may end up harming those around you a bit less.

The science around diet and its impact on the smell of your gas remains underdeveloped at best, but overall, the scientific community is starting to rethink the conventional wisdom that smelly farters should eat less fibre. Although upping your fibre intake may increase the amount of wind you pass, the evidence suggests that it changes the smell of that wind for the better.

The winds of change

If you want to experience the cutting edge of fart-based humour, visit a primary-school playground. Children, connoisseurs of the art, can fart on cue, and the loudest and smelliest will live long in classroom infamy. Anyone who has experienced this environment will know the old rhyme:

Beans, beans, the musical fruit,
the more you eat the more you toot,
the more you toot the better you feel,
so eat your beans for every meal.

There are some foods which are infamous gas producers, and beans do indeed fall into this category. Other notorious gas-producing foods include red meat, vegetables like cauliflower, broccoli, cabbage and Brussels sprouts, as well as garlic, dried apricots, some aromatic spices and even beer.

Broccoli, cabbages, Brussels sprouts and beans are high in a carbohydrate called raffinose, which cannot be digested very well by the upper part of the digestive tract. When raffinose and other poorly digestible sugars reach the colon, the bacteria that inhabit this final part of the digestive tract feed on them, producing gas. Lactose, fructose and sorbitol are all poorly digested sugars, and each of these three substances which can potentially be gas-forming in humans deserves a closer look.

Lactose

Lactose and its 'intolerance' is a tricky beast to get ones head around. Lactose is the main sugar present in milk. To simplify, those who are deficient in the lactase enzyme in their small intestine will essentially not be able to digest the lactose sugar. The failure to digest and/or absorb the lactose sugar in the small intestine is called 'lactose malabsorption'. Lactose intolerance happens because of lactose malabsorption. Lactose intolerance results in symptoms like feeling gassy, then painfully bloated, followed by cramping tummy ache and a lot of flatulence and diarrhea. If you are a bacterium in the colon, getting to feast on lactose is like all your birthdays coming at once, and so when a lot of lactose enters the colon, these bacteria quickly get to work fermenting it, producing various gases. The colon is a pretty well-sealed system, so any gas that these bacteria produce doesn't really have anywhere to go, except for the obvious exit. Therefore, people with lactose intolerance report all those troubling symptoms.

In Northern European countries like the UK, Sweden, Holland, Belgium and Ireland, most people are felt to be able to absorb lactose, partially because of a climate that encourages dairy farming, leading to a wide range of delicious dairy produce that we eat from a very early age.

In East Asian and African communities where milk is not traditionally part of the post-weaning diet, lactase enzyme deficiency can affect nearly 100 per cent of the population: although it has to be said that lactase enzyme deficiency does not necessarily always translate to the troubling symptoms of lactose intolerance in everyone.

Note: it's important to remember that lactose intolerance should only ever be diagnosed by trained medical professionals. Long-term avoidance of dairy can have an impact on health, particularly the bones, since it is rich in calcium and can cultivate a range of other beneficial gut bacteria. Dairy-based products are varied and taste wonderful, and avoiding them unnecessarily is, for me, sacrilegious on culinary grounds and not actually that beneficial in terms of health (unless your doctor tells you otherwise).

So please do go and seek help from your doctor if you feel that dairy products are giving you nasty symptoms.

Fructose

Fructose is a natural sugar, typically found in fruit (fresh and dried) and in particularly high concentrations in pears, honey, grapes, molasses and onions. It tends to be absorbed in the small intestine, but for those with issues absorbing this compound, it can result in feelings of bloating. If it manages to get past the small intestine, it will travel to the colon, where bacteria make quick work of it, releasing a variety of gases which cause bloating, cramping pain and even diarrhoea in the unlucky individual.

But more than the fructose that naturally occurs in fruits and other such healthy treats, it is high fructose corn syrup (HFCS – a substance used extensively in the manufacture of fast food) that can be the troublemaker. There is debate in the scientific community about whether high-fructose diets directly cause diabetes and other nasty issues, but the consensus seems to be that there is probably a fairly strong link between them.

There are two types of HFCS: a mixture of 55 per cent fructose and 45 per cent glucose is used mainly in sugary drinks, ice cream and other frozen desserts, while a mixture of 42 per cent fructose and 58 per cent glucose is used primarily in baked goods, cookies, crackers, tinned fruits, pasta sauces and condiments. You might also find the latter mixture in salad dressings, as well as dairy products like sweetened yoghurts.

But why use fructose? As with all things in life, the answer lies in the accounting departments of the big names of the confectionery industry. High-fructose corn syrup tends to be cheaper than traditional beet sugar (and because corn is far more plentiful than sugar cane its price is usually much more stable), and it also a gives better browning on baked goods and it enhances shelf life.

And if you think that the use of HFCS is limited to soft drinks and baked goods, prepare to be disappointed. You'll find it lurking in condiments, wholewheat bread, fat-free yoghurt, pretty much any breakfast bar, peanut butter and even processed meat slices. So, tread carefully.

Sorbitol

Sorbitol is a sugar occurring naturally in some of my favourite fruits. Blackberries, raspberries, strawberries, apples, cherries, peaches and plums are all rich in this natural sweetener, but while in their natural state sorbitol levels are usually manageable, the concentration increases when these fruits are dried. Horseradish and wasabi are also rich sources of sorbitol, which is surprising as most people wouldn't call wasabi particularly sweet.

The artificial form of sorbitol is usually in 'diet' or 'light' products, listed as the additive E420. Large amounts of sorbitol can cause significant digestive symptoms like bloating, pain and diarrhoea. Sorbitol is used as a sweetener in a vast array of 'sugar-free' products, from diabetic sweets to chewing gum, and it's the reason why some chewing gums have a warning on the label telling you that too much can act as a laxative. Some baked goods also have sorbitol added to them in order to extend their shelf life and to keep the bread moist. Too much sorbitol can have the same effect as lactose intolerance or fructose malabsorption, so it's probably wise to avoid too much of the stuff if you can.

Bread, bloating and coeliac disease

Coeliac disease is a serious illness where the body's immune system attacks its own tissues when exposed to gluten, a protein found in bread and other foods. A normal, healthy gut lining has tiny arms called villi which float around, ready to absorb nutrients into the bloodstream. But coeliac sufferers' villi become flattened and smothered in inflammatory cells until the gluten is removed from their diet, leaving them unable to absorb any other nutrients.

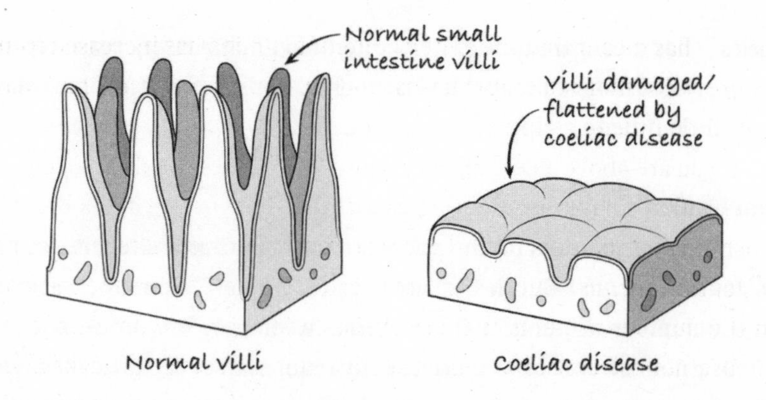

Coeliac disease is different to gluten intolerance, in that rather than the body being deficient in a particular enzyme (an intolerance), gluten triggers an autoimmune response in the body (a disease). Instead of fighting viruses and other nasty stuff like it should do, in coeliac sufferers the body's natural defence system mistakes gluten for a threat and launches a very painful attack on the lining of the gut wall. You can diagnose coeliac disease through blood tests, or by examining samples of the lining of the gut wall under a microscope. You might suspect you have it if you spot excessive flatulence and bloating, cramps, nausea, tiredness, diarrhoea and/or constipation, weight loss, iron and other nutrient deficiencies or even mouth ulcers. Unfortunately, the only way to relieve these symptoms is to completely avoid gluten, which in this day and age is a pretty tall order.

Coeliac disease is not a new phenomenon, and the first descriptions that historians have discovered come from nearly two millennia ago. The Greek physician Aretaeus of Cappadocia used the term coeliac in the first century AD, deriving it from the word *koiliakos* (meaning abdomen in Greek) to describe a patient suffering with an inability to digest wheat. With the advent of agricultural practices, gluten-containing plants took only a few thousand years to spread across the globe (a mere blink of an eye in historical and evolutionary terms). This quick spread – accelerated even further over the last few centuries with the agricultural revolution and the formation of large towns and cities – has meant that the gluten content of grains has increased from its previous, more tolerable levels, and our immune systems have had very little time to adapt.

If you are above a certain age, 'gluten-free' is probably a term that you wouldn't have come across as a child. Gluten-free aisles did not exist in supermarkets, and there were no gluten-free alternatives on restaurant menus. But in the last decade, we have seen an explosion in the number of items that are available for people who cannot (or choose not to) consume gluten. From gluten-free pasta, cakes and bread, to wraps tinted mauve with beetroot and even a beer by Stella Artois, gluten-free is now a multi-million- (some say multi-billion-) pound industry.

The popularity of the gluten-free diet can, at least in part, be credited to its adoption by a number of famous people. Five-time Wimbledon champion Novak Djokovic was one of the first sports-people to go on record with his new gluten-free diet, and it sent some serious ripples through the sporting community. Pop icon Miley Cyrus revealed a 'gluten allergy' in 2012, and famed actress (and gut-health enthusiast) Gwyneth Paltrow published a gluten-free cookbook the following year. The international clothing brand Zara even brought out a t-shirt adorned with the slogan 'Are you gluten free?' Very shortly afterwards, they apologised to the large number of understandably upset coeliac sufferers for cashing in on their problem, and withdrew the item from sale. Even Kourtney Kardashian went gluten free a while back, which no doubt caused a few hundred thousand of her followers to do the same. It seems going gluten free has become something of a cultural phenomenon.

The reality, though, is that while 1 in every 100 people in the UK is thought to suffer from coeliac disease, many of these are undiagnosed. There are also many who make the decision to avoid gluten (mainly bread and pasta) in the absence of any objective evidence that it will do them any harm at all. Those who self-select a gluten-free diet, but have no medical evidence of coeliac disease are termed 'non-coeliac gluten sensitivity' sufferers.

While these non-coeliac gluten sensitivity sufferers report similarly unpleasant symptoms to those with the disease, they don't have the abnormal antibodies on their blood tests, and their villi appear to be perfectly normal under a microscope. Now, these people are most likely not making up their symptoms, so what is at play here? Is non-coeliac gluten sensitivity, as some researchers believe, a type of IBS? Or something else entirely? Or, is there something about the bread we eat these days that might be the cause of these symptoms?

Well, maybe. Bread flour is no longer milled in the same way it was in the past, and the craftsmanship so important in small-scale baking has practically vanished in the age of commercially baked, perfectly formed squares of tasteless white bread. Consider this: the list of ingredients in a loaf of bread that you might have bought a few centuries ago would have comprised five items (water, flour, yeast, sugar, oil), whereas for a generic white loaf today it will include no fewer than *twenty-five* items.

Now, we are not going to make the cardinal mistake here that a synthetic-sounding ingredient must automatically be bad for you, because that in some ways is also a slippery slope to go down. However, it does push us to ask the question: could other molecules similar to gluten, or an additive or an emulsifier in these commercially produced bread products cause the digestive symptoms in people who have non-coeliac gluten sensitivity?

The sad truth is that researchers are nowhere near able to answer this question, but the number of people avoiding bread or other gluten-based products in their diets is increasing. Obviously, people don't need a medical reason to alter their diets and personal preference can be a great motivator, but in my clinical practice, I have found that those who have non-coeliac gluten sensitivity seem to tolerate sourdough bread more easily than, say, a plain white square-cut loaf.

Sourdough is a bread that has been allowed to ferment for many hours, giving it its characteristic taste and texture. Courtesy of the live sourdough starter culture, grains and other proteins will have had the requisite time to break down under the influence of bacteria, making them far easier to digest. Some of my patients have reported switching from modern high-gluten wheat grains to more ancient varieties, such as spelt, with positive results. Right now, with so little actual research on the topic, my advice is that if it makes you feel good and you see positive changes, then by all means go for it – but do so in the knowledge that (unless you have a diagnosis of coeliac or a related disease) the benefits you feel will most likely be lifestyle-based rather than medically based.

Take my (bad) breath away

Some things elicit more burping (or to give it its medical name, gastric eructation) than others; for me, Persian meals consisting of grilled meats and buttery saffron rice are the worst offenders, but once I have had the customary mint tea and an After Eight, the urge to eruct becomes overwhelming.

Burping happens both intentionally and unintentionally. Unintentional, involuntary belching typically occurs after meals to release the air that we swallow with our food, which stretches the stomach. Certain foods, like peppermint, chocolate and fats, relax the ring-shaped sphincter muscle around the lower end of the gullet where it joins the stomach, encouraging the delivery of a satisfying burp. This is probably why mints or mint chocolates are served in restaurants alongside the bill: an attempt to make you more comfortable by releasing any trapped gas that was ingested with your meal.

Alongside burping, other gases have a tendency to emerge from our mouths, and usually at the most inopportune times. Halitosis, the medical term for bad breath, has plagued first kisses and workplace conversations since days of old.

Some foods linger on the breath. Tuna, coffee and garlic come to mind immediately, but it may surprise you to learn that what we eat is not necessarily what causes bad breath. The culprits are actually the billions of microscopic bacteria that make their home around our teeth and gums. Most of their diet is made up of the food we eat, but they also like to feast on the secretions which drip from the back of the nose, down the back of the throat and into the mouth.

Now, while most of us manage to control our mouth odour by brushing our teeth and tongues, there are some unlucky people who still suffer with halitosis, no matter how much they brush. For some, the smell originates in the stomach. Various conditions like diabetes can delay the rate at which the stomach empties, leaving food to ferment and produce pungent gases that rise up the oesophagus and exit via the mouth. But for most people, bad breath is the result of certain unwelcome bacteria taking up residence in the nooks and crannies of the mouth. These bacteria give off particularly smelly gases like hydrogen sulphate (which, as we learned earlier, smells of rotten eggs), methyl mercaptan (which smells of rotting cabbage) and dimethyl sulphide (which smells of rotting seaweed).

And it's not just these sulphurous fumes that cause problems for sufferers of halitosis, even though the compounds mentioned above do make up the majority of bad-breath odour. Scientists have also identi-

Halitosis through the ages

Ancient Egyptians invented a breath mint from a concoction of herbs and spices, and Hippocrates, the father of medicine, allegedly advised young girls to rinse their mouths with wine, dill seed and anise to eliminate those foul oral odours. Both Jewish and Islamic traditions refer to remedies for bad breath by chewing of a mastic gum made from tree resin. But it wasn't until the 15th century that the Chinese invented bristle toothbrushes, using hair from pig necks.

fied close to 150 molecular compounds present in human exhalations which smell, frankly, terrible. For example, the chemicals cadaverine and putrescine (can you spot a pattern in their names?) are present in both rotting dead bodies and bad breath.

There are two families of bacteria that live in our mouths. Above the gum line, sitting happily on our teeth, we tend to find 'gram-positive' species of bacteria which live in our dental plaque. You will have heard about plaque (the living film of bacteria that covers our teeth) from your dentist and also every toothpaste advert that has ever existed. Gram-positive bacteria love sugars, but when they digest them they release acids which break down your tooth enamel.

However, if you want to get to the root of bad breath, go looking for the gram-negative bacteria that live *under* the gum line, in gaps between the teeth and deep in the pits of the tongue. These are the ones that produce those nasty smells. But those gram-positive communities of bacteria are by no means innocent bystanders. Some strains of gram-positive bacteria secrete an enzyme that snips sugar molecules off the proteins found in food, in turn making them more easily digestible for neighbouring gram-negative organisms. The more proteins the gram-negatives digest, the more odours they will release. So, while those gram-negative bacteria are the producers of your bad breath, the gram-positive bacteria are definitely the enablers.

Fresh breath reflects a healthy mouth, but a healthy mouth is not necessarily one that lacks those smelly-gas-forming bacteria completely. Instead, it is one in which the overlapping colonies of bacteria are able to hold one another in check. The ultimate solution to bad breath will eventually be the ability to engineer bacterial communities, seeding more of those that don't generate bad odours, and targeting treatment to get rid of the ones that do. Scientists are currently trying to find a way to do just this, but until then, just stick to the advice of your dentist (and carry a packet of mints and some floss with you).

Irritable bowel syndrome

If you suffer from IBS, you are part of a very well-populated club. Along with you and one in every ten other people in the world, notable figures like John F. Kennedy, Tyra Banks, Kirsten Dunst and Marilyn Monroe have all belonged to this club.

Irritable bowel syndrome was first mentioned in 1849 by an otherwise unknown Dr Cumming in the *London Medical Gazette* as a condition where 'the bowels are at one time constipated, another lax, in the same person'. He also added to this the very helpful statement: 'How the disease has two such different symptoms, I do not profess to explain.' So, thanks for that, doc.

Today, we know that women seem to be diagnosed more than men, and most people are affected in the third or fourth decade of their lives. Living with IBS is often a constant battle that leaves sufferers feeling both exhausted and frustrated; it can affect every facet of daily life, from relationships and social life, to school and work performance (and the ability to sit through a Thai massage without gassing the entire room – see pp. 252–254).

One of the key features of this syndrome is that there are no consistent structural, biochemical or radiological issues that can be shown to cause it. This often leaves sufferers a little annoyed, because (of course) they want to know the underlying reason for their discomfort – a physiological issue that is causing them to feel the way they do. As a gastroenterologist, it is heartbreaking to have to tell very symptomatic people that all their tests are, well, normal.

Officially speaking, IBS is a diagnosis you can only make when other sinister causes for symptoms have been excluded. This means that if you go to your GP with IBS-like symptoms, a number of investigations will follow, particularly if you are above a certain age. If, once these investigations are completed, nothing definitive has been found, you might be diagnosed with one of the three subtypes of IBS: IBS-D (where diarrhoea is the main symptom), IBS-C (where constipation is the main symptom) or IBS-M (mixed – if you alternate between the two).

Regardless of which letter ends up being attached to your diagnosis, bloating is the common thread, reported by 96 per cent of patients with IBS. Doctors look at three main diagnostic criteria to check for IBS (pain related to starting a poo; changes in the frequency of poos; and changes in the appearance of poos), but bloating – despite not featuring in the diagnostic criteria – is often considered by many to be the most bothersome symptom. In fact, the girth of the abdomen can increase by up to *12 centimetres* in some IBS patients.

Around two-thirds of IBS sufferers directly attribute their symptoms to what they are eating. They report spending their time trialling various diets of exclusion in the hope of pinpointing the causes of their discomfort, which can have a huge impact on their ability to live a normal life. Imagine every time you faced food, be it breakfast, lunch, dinner or even a little snack, having to ask yourself the question: will I regret eating this? Will whatever I am eating now make me ill in a few hours? From the viewpoint of a very avid eater, this affliction sounds like torture. In fact, if you read the personal accounts from IBS sufferers on the many online IBS support blogs, you can get an idea of just how disabling the condition can be. Here is one anonymously posted testimonial from a sixteen-year-old:

> *Each morning, I awake feeling nauseous. This nausea usually lasts for several hours, sometimes the whole day. This is also accompanied with extreme abdominal pain and diarrhoea. I often do not eat a lot, usually because I have no appetite or I am afraid of a flare-up. My flare-ups happen quite often and at random times, even if I haven't eaten anything to cause one ... I am often not able to do the same things that other girls are able to do my age, and I am quite miserable with having to live like this day in and day out.*

So, what do we know so far about why people get the condition? We know that if you have an affected parent, the chances of you having

IBS are much higher. We know that it makes the way your food moves through your digestive system very disturbed and irregular. We know that IBS sufferers demonstrate volatile gut microbiomes, which, as we saw in the previous chapter, is extremely important for maintaining gut health. We know that symptoms are far worse when sufferers are living through periods of stress, as this can cause the release of hormones like cortisol and adrenaline and this response might be exaggerated in people with IBS.

We also know, adding insult to injury, that IBS patients can demonstrate a fascinating phenomenon called 'visceral hypersensitivity', which is an increase in the sensation of pain experienced from internal organs. So not only does IBS hurt, but each painful episode is amplified compared to how it might feel in a non-IBS sufferer's body. Talk about kicking them when they're down.

What can I eat to improve IBS?

There is actually very little information for IBS sufferers on foods that might help, which is frustrating for them; but it makes sense when you consider that IBS doesn't really have a defined cause. The key messages, according to a dietician colleague, are very simple:

› Cook for yourself, if and when you can, and try to eat three regular meals a day.

› Avoid processed foods that can irritate your stomach. Instead of buying a burger, for instance, cook it from scratch at home to avoid unnecessary additives that could cause a flare-up.

› Eat plenty of fruit, veg, nuts and seeds.

› Limit alcohol intake, particularly binges, as alcohol affects how effective the intestines are.

- Cut down on caffeine and artificial sweeteners (like those found in diet drinks and chewing gum).

- If your IBS symptoms include bloating, avoid gas-forming foods like beans, Brussels sprouts, cauliflower. Oats and mint tea can be beneficial.

- If your IBS symptoms include constipation, then a *gradual* increase in fibre consumption may help (but be aware, sudden increases can make symptoms a lot worse).

- At least to begin with, avoid trying lactose-free or gluten-free diets, unless instructed by your doctor. There's very little evidence to suggest that these diets will lessen IBS symptoms.

- Try a probiotic supplement and monitor its effects. If that doesn't work, try a different one after you end the first trial. Note: each probiotic that is trialled needs to be continued without a break for at least a month at a time.

- If you have suffered from bloating, abdominal pain and diarrhoea for more than a month, try (with your dietician) a low-FODMAP diet (see overleaf). If, after a month of strict adherence, no benefit is observed, then the diet should be stopped.

And, that's it. People are always looking for a magic ingredient, a superfood or new diet that revolutionises their life completely, but the dietary messages for managing IBS are mostly based on common sense. Bottles of rancid apple-cider vinegar, cupboards full of goji berries and herbal teas that taste of nothing but cost £2 per bag are, sadly, not going to change things much.

FODMAP

FODMAP is an acronym for fermentable oligosaccharides, disaccharides, monosaccharides and polyols, which are short-chain carbohydrates that are resistant to digestion. Instead of being absorbed by the small bowel into the bloodstream, they reach the large intestine where most of the gut bacteria reside. There, they are fermented to produce gas, and can also cause diarrhoea and abdominal discomfort in people suffering with irritable bowel syndrome. In theory, the fewer of these there are in your gut, the less likely you might be to suffer from symptoms of IBS.

I have been extremely wary of including information on the low-FODMAP diet for a few reasons. It has been used to alleviate suffering for people with severe IBS, but anyone who goes on to one should be doing so under the supervision of a registered dietician according to our national guidelines. In clinical practice, this sort of diet is reserved mostly for IBS patients with severe symptoms, rather than the population at large, because improper use can deprive the body of essential nutrients. The low-FODMAP diet is also highly restrictive and is designed for use in the short term only. So, please don't dive into anything without consulting your doctor or dietician.

For all these reasons, it is a source of frustration for me to see some authors encouraging the use of a low-FODMAP diet in their books as a cure for bloating. But – *and I cannot stress this enough* – it is not that simple.

None the less, I have created a few recipes that are both FODMAP friendly for those with IBS observing the diet, but also perfectly delicious for people who don't have IBS but find themselves occasionally bloated – or for anyone at all.

- **Fermentable:** gut bacteria digest these carbohydrates to produce gases.
- **Oligosaccharides:** fructans and galacto-oligosaccharides e.g. vegetables such as artichokes, asparagus, Brussels sprouts, broccoli, beetroot, garlic and onions; grains such as wheat and rye; chickpeas, lentils, kidney beans and soy products.
- **Disaccharides:** lactose found in dairy products like milk, soft cheese, yoghurt, ice cream.
- **Monosaccharides:** fructose as found in honey, agave nectar, apples, pears, peaches, cherries, mangoes, pears and watermelon and products made with high-fructose corn syrup.
- **Polyols**, namely sorbitol and mannitol, found in: fruits such as apples, apricots, blackberries, cherries, nectarines, pears, peaches, plums and watermelon; vegetables such as cauliflower, mushrooms and snow peas; sweeteners such as sorbitol, mannitol, xylitol, maltitol and isomalt (found in sugar-free gum and mints) and cough medicines and drops.

Food Category	High FODMAP foods	Low FODMAP foods
Vegetables	Artichoke, asparagus, cauliflower, garlic, green peas, leek, mushrooms, onion, sugar snap peas	Aubergine, beans (green), bok choi, capsicum, courgette, lettuce, potato, tomato
Fruits	Apples, apple juice, cherries, dried fruit, mango, nectarines, peaches, pears, plums, watermelon	Cantaloupe, grapes, kiwi fruit, mandarin, orange, pineapple, strawberries
Dairy alternatives	Cow's milk, custard, evaporated milk, ice cream, soy milk (made from whole soy beans) sweetened condensed milk, yoghurt	Almond milk, brie / camembert cheese, feta cheese, hard cheeses, lactose-free milk, soy milk (made from soy protein)
Protein sources	Most legumes / pulses, some marinated meats / poultry / seafood, some processed meats	Eggs, firm tofu, plain cooked meats / poultry / seafood, tempeh
Breads and cereal products	Wheat / rye / barley based breads, breakfast cereals, biscuits and snack products	Corn flakes, oats, quinoa flakes, quinoa / rice / corn pasta, rice cakes (plain), sourdough spelt bread, wheat, rye, barely breads
Sugars, sweeteners and confectionary	High fructose corn syrup, honey, sugar free confectionery	Dark chocolate, maple syrup, rice malt syrup, table sugar
Nuts and seeds	Cashews, pistachios	Macadamias, peanuts, pumpkin seeds, walnuts

Summary

- Reducing the stigma around bloating and starting the conversation about why we fart and what our wind is composed of is essential.

- For the most part it is sulphurous compounds that produce eggy-smelling farts, and these also contribute to bad breath, or halitosis.

- There is a suggestion that 'resistant starches' may help with noxious farts, and that reducing fibre will only decrease their frequency, not their potency.

- Rates of lactose intolerance vary widely according to geography. Avoiding dairy without medical input can have long-term health consequences.

- Fructose (as found in processed foods containing high-fructose corn syrup) and sorbitol (found in artificially sweetened products) can result in bloating and other bowel symptoms.

- Coeliac disease is an autoimmune condition that many people don't realise may be the cause of their debilitating digestive issues. Conversely, some people who don't actually have coeliac disease avoid gluten and tend to report some benefit.

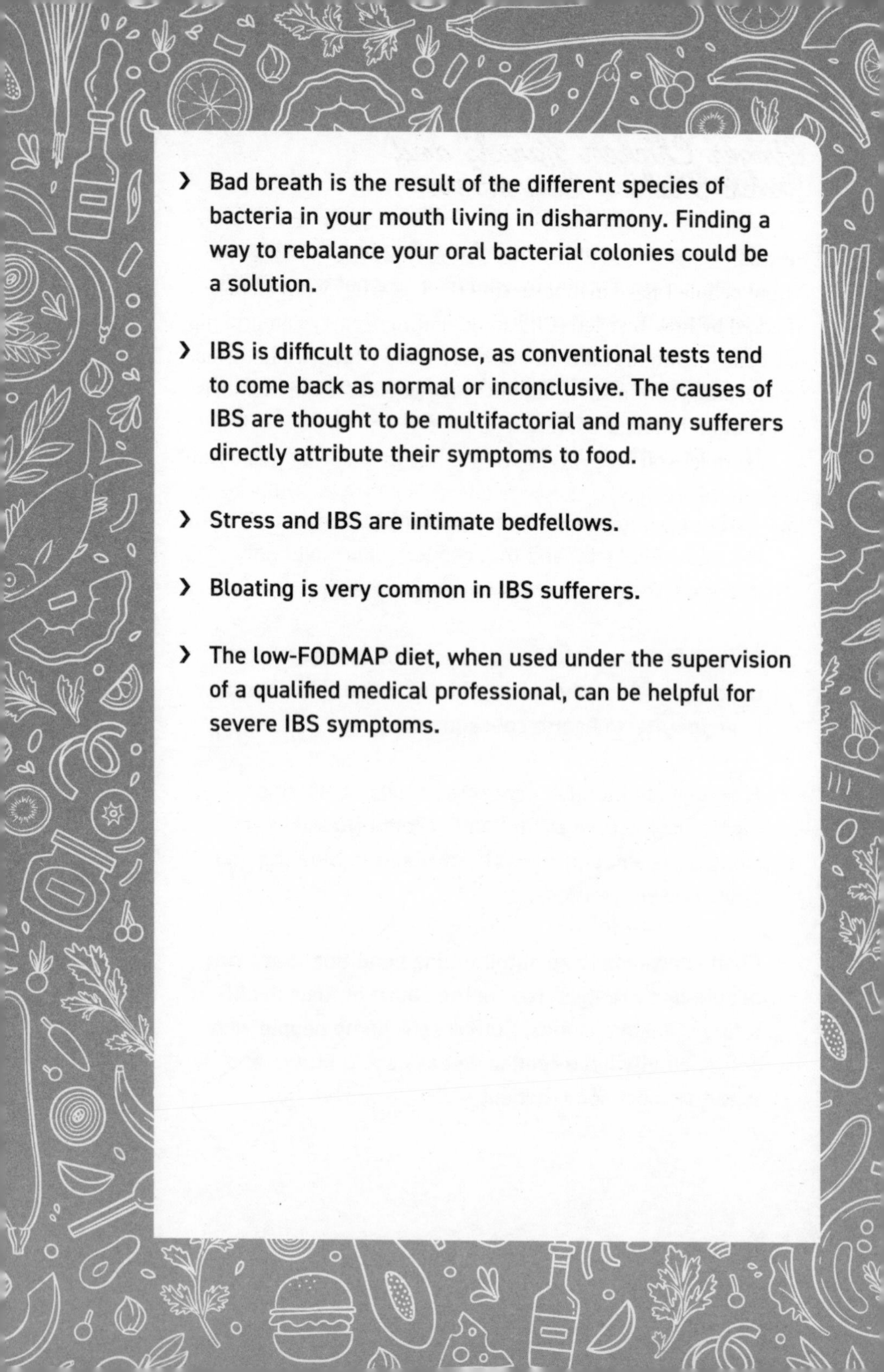

- Bad breath is the result of the different species of bacteria in your mouth living in disharmony. Finding a way to rebalance your oral bacterial colonies could be a solution.
- IBS is difficult to diagnose, as conventional tests tend to come back as normal or inconclusive. The causes of IBS are thought to be multifactorial and many sufferers directly attribute their symptoms to food.
- Stress and IBS are intimate bedfellows.
- Bloating is very common in IBS sufferers.
- The low-FODMAP diet, when used under the supervision of a qualified medical professional, can be helpful for severe IBS symptoms.

Ginger Chicken Karahi and Quick Pickled Cucumbers

Serves 2

Many people I have met observing the low-FODMAP diet are affected by how restrictive it can be. Eating a curry can be a nightmare, as it contains both onions and garlic (which are notoriously gas-forming). Even shop-bought curry powder and garam masala can contain hidden FODMAPs.

Just to let you know, I use the green tips of spring onions and leeks in this recipe, as these are low in FODMAPs, while avoiding the roots of both (which are notoriously high in them).

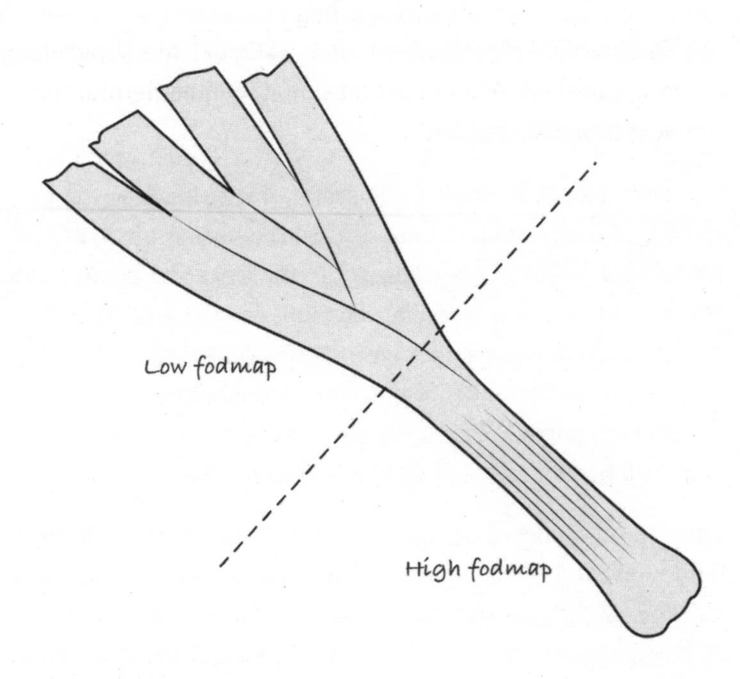

Ingredients

400g leek tips
4 spring onions, green tips only
4 tablespoons vegetable oil
75g ginger
½ teaspoon turmeric
¼ teaspoon asafoetida
1 teaspoon red chilli powder
1 teaspoon ground coriander
5 tomatoes, roughly chopped
500g boneless chicken thighs, diced
Salt, to taste
A handful of coriander, chopped

For the cucumber pickle:
1 medium-sized Turkish cucumber (or ½ a normal cucumber)
½ teaspoon salt
½ teaspoon sugar
2 teaspoons white vinegar

Method

1. Finely slice the leek and spring onion tips. Heat the vegetable oil in a shallow non-stick casserole dish and fry the leek and spring onion tips over a medium heat for around 5 minutes until they soften and brown slightly.

2. Grate the ginger through a fine grater. This will allow you to extract the pulp and juice and discard the fibrous bits that get left behind. Add the grated ginger to the leeks and onions, along with the turmeric, asafoetida, red chilli powder and ground coriander. When the spices have toasted off for around a minute or so in the pan, add the roughly chopped tomatoes. Take care not to allow the spices to burn. Stir everything together thoroughly to help it all break down and combine to form a 'masala'.

3. Once the contents of the pan have mostly dried out (around 5 minutes) and the oil appears to be splitting out, add the diced chicken thighs. Stir the chicken into the masala, and continue to do so regularly over the next 15 minutes or so; the contents of the pan will initially look quite dry, but resist the temptation to add extra water as the chicken will release lots of water as it cooks.

In total, the chicken needs about 25 minutes of cooking in the masala to soften fully.

4. The dish is ready when most of the moisture created by the chicken has evaporated and once again the oil starts splitting out of the masala. At this stage, season with salt to taste. Garnish with the coriander before serving.

5. For the cucumber pickle: use a potato peeler to slice the cucumber lengthways into ribbons and discard the seeded, watery centre. Mix the cucumber with the salt, sugar and vinegar and rub them gently with your fingertips into the cucumber. Serve alongside the Ginger Chicken Karahi with some boiled rice and gluten-free flatbreads of your choice.

Baked Potato Rosti with Cheddar and Za'atar

Serves 4

Not all cheeses are created equal when it comes to lactose. As a rule of thumb, it is the ripened ones that contain less of this sugar. It is partly to do with the way they are produced, where the whey is discarded and also because bacteria that contain the enzyme lactase will have chipped away at any remaining lactose in the cheese. Blue cheese, brie and Camembert are all considered low FODMAP in controlled quantities. A 40-gram portion of Cheddar is low FODMAP, making this rosti-style potato cake fantastic as picnic fare, brunch or a side.

Ingredients

1.25kg waxy potatoes
20g chives, finely chopped
1 tablespoon gluten-free flour
1 egg
Vegetable oil, to grease
1 heaped teaspoon za'atar

135g extra strong Cheddar cheese, grated

Sea salt flakes

Method

1. Preheat the oven to 180°C fan.
2. Grate the potatoes and sprinkle liberally with salt. Set aside for 15 minutes before squeezing as much of the water out of the potatoes as possible. You do have to squeeze quite hard; the more liquid you squeeze out, the crispier the final rosti cake will be. A little tip is to place the potatoes in batches in a clean J-cloth and wring to squeeze out the moisture.
3. Add the chives, flour and egg to the grated potato and mix well.
4. Now grease a 23cm greaseproof cake tin liberally with vegetable oil. Pour in half the potato mixture and pat it gently to cover the base of the tin. Try not to press down too firmly as you run the risk of making a dense potato cake otherwise. Now sprinkle all but a handful of the cheese on to the first layer of potato, followed by the za'atar. Complete the cake by layering the remaining potato over the cheese. Again, try not to press the potato down too firmly. Top with the remaining cheese, cover loosely with a piece of foil and transfer to the oven. Bake for 25 minutes, then remove the foil and bake uncovered for another 25 minutes until the rosti cake looks dark golden and crisp. Eat while still hot.

The No-bloat Salad

Serves 4–6

Many people who suffer with bloating resort to excluding salad from their diets on the assumption that it will worsen their symptoms. We know, however, that some salad leaves are low in FODMAPs while others are high. For example, white and red cabbage are actually low FODMAP, despite other varieties being high. So, my point is this: not all salads are created equal, and just because you bloat, it doesn't mean that you need to exclude all salad from your diet.

This low-FODMAP salad is paired with a spicy dressing, but you could just as well use a low-FODMAP mayonnaise or other citrusy dressing of your choice.

Ingredients

For the dressing:

25g fresh ginger
1 teaspoon paprika powder
½–1 red chilli (depending on how hot you prefer it)
1 teaspoon cumin seeds
1 teaspoon coriander seeds
Juice of 2–3 limes
1 tablespoon fish sauce
2 tablespoons maple syrup
3 tablespoons olive oil
Salt, to taste

300g white cabbage, finely shredded
200g tomatoes, sliced
75g green beans, finely sliced in half lengthways with a sharp knife
3 medium-sized carrots, finely shredded
150g beansprouts
Handful dry roasted peanuts

Method

1. Combine all the ingredients for the dressing in a blender with approximately 75ml warm water. Blitz to a fine purée. Season the dressing with salt to taste and set aside. You can add a touch more water if you want to thin it down further.

2. Now place the cabbage, tomatoes, green beans, carrots and beansprouts in a bowl. Drizzle over the spicy dressing and toss everything together to ensure that all the vegetables are coated evenly. Taste, and add extra salt if you wish. Top the salad with crunchy peanuts before serving.

Melon, Lime and Rose Sorbet

Serves 6

A rather sophisticated fruity dessert, with the added benefit of being low FODMAP, dispelling the myth that because fruits contain the sugar 'fructose', they will all be high in FODMAPs. Cantaloupe, honeydew and Galia melons, unripe bananas, blueberries, kiwis, passion fruit, pineapple, papayas, raspberries, strawberries, lemons, limes and oranges are all low in FODMAPs and therefore unlikely to cause bloating.

Ingredients

700g ripe Galia melon flesh, roughly chopped (discard the seeds and skin)
Juice of 4 large or 6 small limes and zest of 3
150g icing sugar
2 tablespoons rosewater

Method

1. Place the melon, lime juice and zest, with the icing sugar in a blender and blitz to a very smooth purée. I use a NutriBullet, which gives a very smooth finish. Add the rosewater to the blended mixture and mix gently with a whisk before churning in an ice-cream machine for around an hour (follow the manufacturer's instructions).

2. If you do not own an ice-cream machine, you can still make this sorbet, but it will require a little more work. You will need to remove the sorbet from the freezer every 30 minutes for 4 hours and beat it with an electric whisk to prevent ice crystals from forming.
3. Leave the sorbet to stand (around 10 minutes for the churned version; otherwise up to 30 minutes) before scooping to serve.

I hope that I have not laid on the science too thick here because this, after all, is a light read about food and not a gastroenterology textbook. None the less, after reading this chapter you will appreciate that bloating and flatulence are complex phenomena, influenced by many foods/ingredients.

Understanding the processes that take place within our bodies, especially those that are triggered by certain foods, can help any of you readers who, like my friend Delia, might have been suffering with these symptoms but weren't sure how to manage them. It may also help those who simply want to know more about how our bodies work, as the reality is that we will all suffer with bloating and excessive flatulence at one time or another in life – it is perfectly normal.

Your digestive system speaks to you all day, sending symptoms from your body to your conscious perception. Part of respecting your body is listening to it and being in tune with any digestive symptoms that may be experienced. Reading the physical cues when you find yourself bloated and making food choices that honour your health, while concurrently satisfying the taste buds is vital. Being accepting of your wind and fostering a deeper understanding of how food can affect it is yet one step more in the quest to achieve digestive health and happiness.

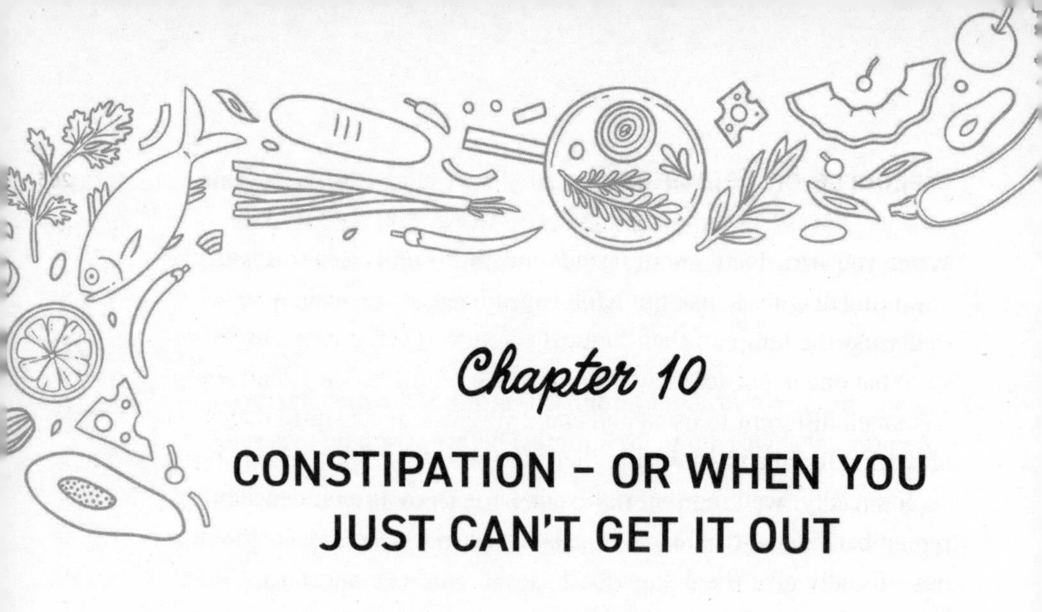

Chapter 10

CONSTIPATION – OR WHEN YOU JUST CAN'T GET IT OUT

Maybe this will resonate with some of my readers, or maybe you'll all simply finish this book knowing just how weird my brain is, but when I am constipated, I find myself doing something rather peculiar. I start to reminisce about all those times I could poo comfortably and freely, and how I never really appreciated them for the magical experience that they were.

This chapter, like the preceding one, focuses on things in the body not going quite to plan. Previously, we looked at bloating, and what happens when gas builds up inside us; now, we will take a look at constipation – because as much as we don't really want to be reminded of it, our poo is an intrinsic part of our relationship with food, and the final stretch of the digestive system can make or break that relationship, either consolidating or demolishing a sense of digestive health and happiness. Because of this, I think it would be wise to explore the inner workings of the bowel (and hopefully this won't put you off your dinner too much). A word of warning: there is (obviously) going to be a lot of talk about poo.

Discourse on defecation

When you are a doctor, your friends and family will often treat you like a walking diagnosis machine. You can sniff out the cousin who wants to discuss the lump on their bum, the auntie covered in moles who is sure that one in particular wasn't there last month or the friend whose feet smell different to usual and could you just take a sniff. They start off with normal chit-chat, and then the tone of conversation changes.

It's tricky, walking that tightrope between common decency and remembering that the people in question are not actually your patients, but I usually give them some basic advice and then encourage them strongly to consult their GPs. But there are times when even the most diplomatic exit strategy just doesn't work.

It was my friend Hannah's birthday party, which was really an excuse to introduce us all to her new boyfriend, Oscar. Oscar seemed a lovely chap and the party was in full swing, with everybody enjoying the merriment. Then Hannah suddenly grabbed my arm and directed me to the corner of the room.

'Oscar has bowel issues,' she said.

I could see where this was going. 'Tell him to see his GP,' I said. 'I really don't want to know about your boyfriend's bum trouble.'

'Sorry, you haven't got a choice,' said Hannah, with a genuine look of anguish. 'We tried to speak to his GP but got nowhere. I need your help.'

I had a sudden sinking feeling, as Oscar came and sat on the chair across from me, and it looked like he did, too. A handsome chap with a chiselled face and kind demeanour – the sort of man most women would swoon over – Oscar was now being forced by his new girlfriend to discuss his bowel issues with a stranger who happened to be a doctor, in a room full of people he had never met before. He sat, rather uncomfortably, on the edge of his seat, legs crossed, his gaze focused on a spot somewhere on the ground to the left of me and his arms firmly gripping his abdomen. His fists were clenched tight and one foot tapped the floor rhythmically.

Oscar struggled ... 'Basically, when I do my business ... you know ... the other-end stuff ... the nature of it has changed.' The awkwardness of the situation was not lost on either of us, though perhaps it was on Hannah, who was on her fifth Prosecco and was clearly getting impatient with his reticence.

'Just bloody tell her that you only poo once a fortnight and it all comes out looking like Rolos,' she said, tired of waiting for him to tell me himself. She turned to me. 'He doubles up in pain on the toilet seat; I saw him the other day. I'm worried sick.'

Oscar's forehead was flushing bright red. I was about to get off my seat, unable to take any more awkwardness, when Oscar looked at Hannah with a wry smile, before shouting, 'Thank you, Hannah, for providing the executive summary of my problems.' His attention then turned to me. 'And Saliha, since we're clearly doing this, she's right. Lately, it has been pretty difficult to shit, and if and when I can go, it comes out looking like rabbit droppings.'

As someone who worked in an office, Oscar spent up to sixteen hours a day stuck to his chair in an air-conditioned room. There was little time left for exercise, cooking or even water breaks and his meals were usually takeaways. He wanted a way out that didn't involve the nausea and wind that his current mix of laxatives gave him. In the end, we had a pretty long chat about the foods that, going forward, might help him.

This conversation with Hannah and Oscar transformed the way I thought about constipation in many ways. I realised just how embarrassed people can feel talking about the act of defecation, and that, as a society, we have a responsibility to the Oscars around us to remove that feeling of shame. Age-old attitudes need to be put aside; defecation is a normal bodily function and deferring the conversation on our bowel movements will only mean a delay in getting the help we might one day need.

I have also realised that many people would prefer a change in lifestyle or diet to treat constipation, as opposed to taking pills or laxatives, but not many know what those changes should be. Even talking about poo and food in the same sentence can feel uncomfortable for some, and the Internet – our most popular information source – can exacerbate the issue by providing a hundred different viewpoints from a hundred different people.

To make the most of this chapter, I hope that you will be able to leave any hang-ups about poo at the door and allow yourself to take a scientific and culinary journey with me. If, after reading this chapter, you discover that you are able (or at least accept that it is ok) to talk freely about bowel function and food together, then that would be the best gift you could give me. Many people use the term loosely, but in this chapter, I want to get stuck into what it really means to be constipated. I will also chat a little about the sources and benefits of dietary fibre and offer you some convenient (and hopefully enjoyable) ways that you can incorporate more of it into your diet.

When poo won't go through, no matter what you do

At any point in time, around 14 per cent of the world's population are constipated (that's every seventh person you pass when you walk down the high street). Constipation is more common in women and older adults, but, as Oscar will tell you, men are also affected, as are children and babies. The data is a little scant, but it seems you are more likely to be constipated if you live in the Western world (this includes Australia), compared to Asian, African or South American communities.

As a concept, constipation means many different things to different people. This makes defining the issue difficult. For example, does being constipated mean that you pass hard, lumpy poos? Or is it that your poos are the same as before, but less frequent? Are you constipated if it always feels like there is still some poo left to come out, no matter how hard you strain? Do bloating and tummy pain feature; or are they separate issues?

As a gastroenterologist, this ambiguity means that when a patient tells me they are constipated, I can't really do much until they are willing to engage with me in a frank conversation about their poo. I need to know all the details: from texture to colour, odour to taste (OK, not taste). I need to know how they define constipation, and how it impacts their life. People also mistake constipation for a diagnosis,

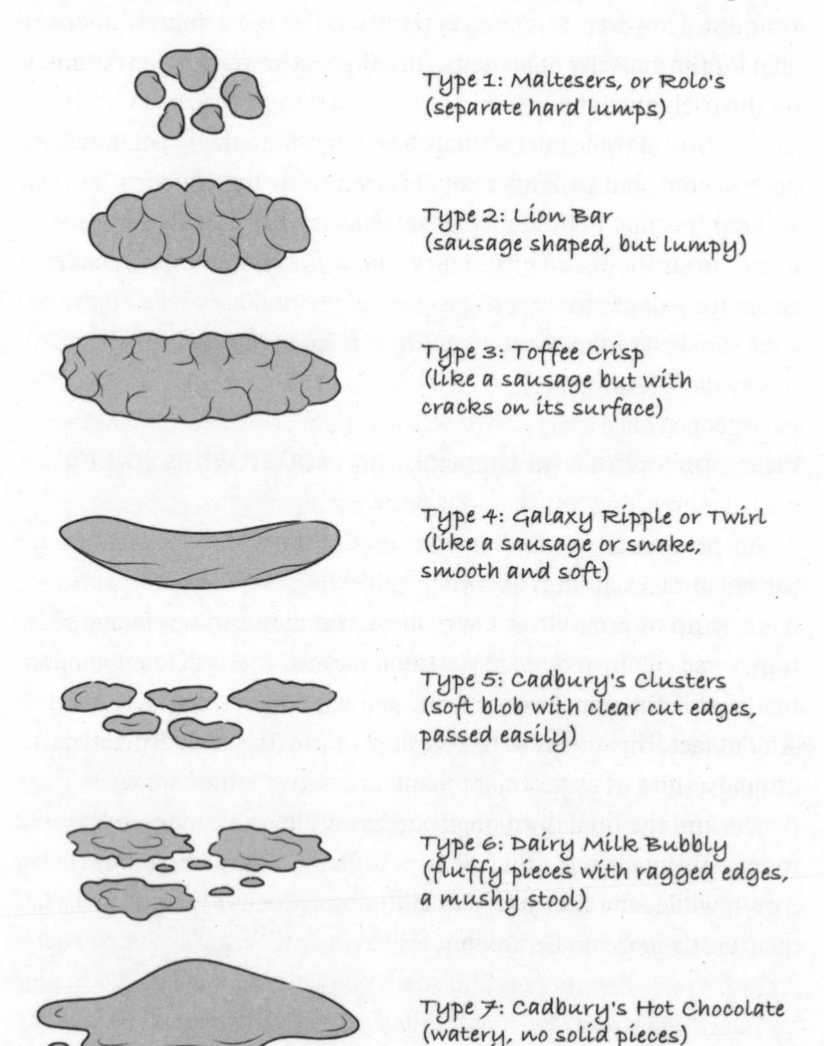

when in reality it is usually a symptom of some other issue that needs to be identified and understood by a clinician.

One thing that has been extremely useful in assisting with the description of stool consistency is the Bristol Stool Chart, a fantastic visual tool to start the conversation. The original version has just the pictures of stool types from 1 to 7, with 1 being the firmest and 7 the runniest. However, this chart is fairly sterile, and definitely not relatable for the majority of patients. The alternative version here seems to do the trick much better:

The Bristol (/chocolate) stool chart is a good starting point, helping both doctors and patients to understand what the patient's 'normal' is. I say 'normal', but in actual fact, this is not really a thing when it comes to our bowels. For instance, the assumption that we should all poo once a day is inherently flawed, as is the theory that everyone's stool should be either Bristol Type 3 or 4 (Toffee Crisp or Ripple, to use the medical terminology). I have met many people who, for decades, have pooped once every five days, so for them, being constipated would mean even less frequent than this; while others, who poo twice a day, consider anything less than this to be outside of their normal.

So, how much should you be pooing? The passage of stool through the colon takes around twenty to thirty hours for most of us, which is made up of around six hours in the right colon, ten hours in the transverse colon, and ten to sixteen hours in the left colon. Doing the maths, this amounts to around one poo a day on average (give or take). And in fact, this seems to be reflected in the data; it is estimated that around a third of us poo more than once a day, a third are once-a-day pooers and the final third open our bowels less than once a day. And for anyone interested, around half of us have poo that resembles Bristol 3 or 4, while for a quarter it's usually a bit more lumpy and for the last quarter it's more on the loose side.

How is poo made?

I've given you some background on constipation, but this book is called *Foodology*, not *Poo-ology*, and we need to know how constipation relates to the food we eat. So, what happens between the mouth and the toilet?

Let's say you've just sat down to a nice bowl of salad. When you eat the salad, you put in motion an exquisitely intricate process. We have already gone over in a lot of detail in previous chapters what happens in the mouth and the stomach, so let's skip those two steps now and move straight to the bowels. The salad (transformed by your stomach into a movable mush) is pushed forward by the walls of your small intestine through a series of wave-like muscle contractions called peristalsis. This is a sophisticated process that is responsible for keeping things moving in the same direction. You can picture it as being a bit like how an earthworm moves itself along a garden path.

We do see some peristalsis movements in the large bowel (the colon), just like those in the oesophagus and small bowel. However, peristalsis on its own is not enough to move faecal material through the entirety of the large bowel. Two other colonic movements – low amplitude propagating contraction (LAPC) and high amplitude propagating contractions (HAPC) – are needed to move our poo through.

LAPC happen about a hundred times a day without us even knowing about them; they serve to move fluid through the colon and assist with the slow passage of wind. In contrast, HAPC are large, high-pressure colonic movements that happen far less frequently, perhaps only six times a day. It is HAPC that push poo into the rectum and give us the urge to run to the loo. This 'urge' (unless you are extremely in tune with your body) is the first time that the movement of poo through the body enters our conscious perception (up to this point it was almost exclusively under our subconscious control) and once we become aware of it, a range of specialised nerves fire into action.

The net result of all these activated nerves is that our bodies generate pressure that makes us feel like we need to poo: the muscular floor of the pelvis relaxes downwards and a muscle called the pubo-rectalis muscle, which normally slings around the rectum and keeps it hitched up tight, loosens, allowing us to expel faeces. When we are all done, the whole process reverses: the pelvic floor rises up, our abdominal muscles relax and the pubo-rectalis muscle slings back up, preventing any leaks or skid marks between bowel motions.

Note: If you experience any change in bowel function from your normal (diarrhoea or constipation), or notice blood or mucus through the back passage, pain on evacuating stool, unexplained tiredness and weight loss, a pain or lump in the tummy, please consult your doctor immediately, as these symptoms can be a warning sign that more sinister things are going on.

Something to talk about

It seems a bit odd that although constipation is something that we can address quite effectively, it is also a subject that many people can feel painfully shy discussing. Have you ever seen those daytime TV adverts targeted at women, featuring the beautiful, successful businesswoman getting stomach cramps while out at lunch with her friends? Why does she have to hide it? Why can't she just say to her smiling, perfect friends, 'Sorry, girls, my belly is doing that thing again. I'm off for a quick poo. Be right back!'

And the problem is not by any means exclusive to women. Half of people, according to one study, cannot poo unless in complete privacy. Some even have trouble pooing if they know someone else is in the house, even if they are alone in a locked toilet. (In contrast, others are of the belief that you are not truly close to someone unless you can poo in front of them.)

There is a name for this affliction: parcopresis, which is the psychologically triggered physical inability to poo without a certain level of privacy. Sufferers of parcopresis are a step above those who are simply

Is all constipation the same?

For those who are constipated and where no worrying pathology has been identified by healthcare professionals (which will be most people), constipation will broadly fall into three sub-groups:

1. Normal transit constipation: this is the most common form of constipation and, sadly, it is not clear why it happens. The time it takes for poo to transit through the bowel is totally normal. However, it is thought that changes in the colon's microbiota, as well as dietary, lifestyle, psychological and behavioural factors, probably all play a part in stools becoming hard and pellet-like where they were previously soft. Fortunately, this type of constipation responds well to dietary changes, particularly a graduated increase in fibre intake.

2. Slow transit constipation: normal peristaltic movements through the colon do not happen, neither do the high-frequency propagation movements mentioned above. The net result is delayed transit time of stool through the colon. If you look at cell specimens of people with slow transit constipation under the microscope, they sometimes have reductions in a type of cell called the 'interstitial cells of Cajal' that are responsible for communication between the muscle in the walls of the bowel and the nervous system.

3. Dysynergic constipation: as you now know, in order to evacuate the bowels, coordination of a number of muscles via nerves is required. In this form of constipation, there are abnormalities of the contraction of the rectum and inadequate relaxation of the anus, leading to impaired evacuation and expulsion of stool.

a bit embarrassed about pooing in public, their bodies being physically unable to relax the necessary muscles when they are not in their familiar toilet setting.

Parcopresis sufferers can sometimes overcome their affliction by desensitising themselves to the societal stigma of poo. Our old friend Oscar beat his by watching the 'Embarrassing Poo Stories' channel on YouTube with Hannah, in an attempt to normalise his perception of the process. I would recommend a watch if listening to other people recount their most mortifying personal experiences is your cup of tea. As I found out when researching this book, it most definitely is mine.

What to do to make you poo

Modern life is, unfortunately, a very powerful driver of constipation. Many people have sedentary jobs, sitting for long stretches, often hours at a time, without moving or stretching. Bowels have evolved inside bodies that roamed, hunted and foraged, and as such, they need to jiggle around in order to move their contents forward. Office work means that they don't get to jiggle as much as we might want them to, making our digestion seem sluggish, backed up. In fact, during the period of lockdown due to the coronavirus pandemic (as I mentioned in a previous chapter, if you are reading this in the future, 2020 was *not* humanity's best year), many people who had previously enjoyed regular bowel movements found themselves surprised at how constipated they had become due to their inability to exercise, play sport or even leave the house.

As it turns out, we don't even poo in the proper position any more. If you have ever travelled around India, you might have been surprised to find that toilets come in two varieties: Western and Eastern. Western ones are easily recognisable, while Eastern ones are simply a porcelain hole in the ground with places either side of the squatting pan to place your feet. In anatomical terms, the optimal position for a poo is to squat (not sit) with the knees positioned at a less than 90-degree angle to the waist, meaning that for comfortable relief of bowels, you should opt for the Eastern toilet every time.

Now, Eastern toilets require some getting used to, and also a not-unsubstantial reserve of thigh strength and stamina. It's not really a position that lends itself to reading or phone scrolling. However, if you are interested in a new poo position, but would still like to check your phone on the loo, consider investing in a Squatty Potty (or any 6-inch-high footstool, or even a couple of stacked books), which helps to raise your feet, improving the angle between your hips and knees, so helping your poos to come out more easily.

But apart from exercise, the Squatty Potty and addressing our societal stigmas around pooing, one of the ways we can address consti-

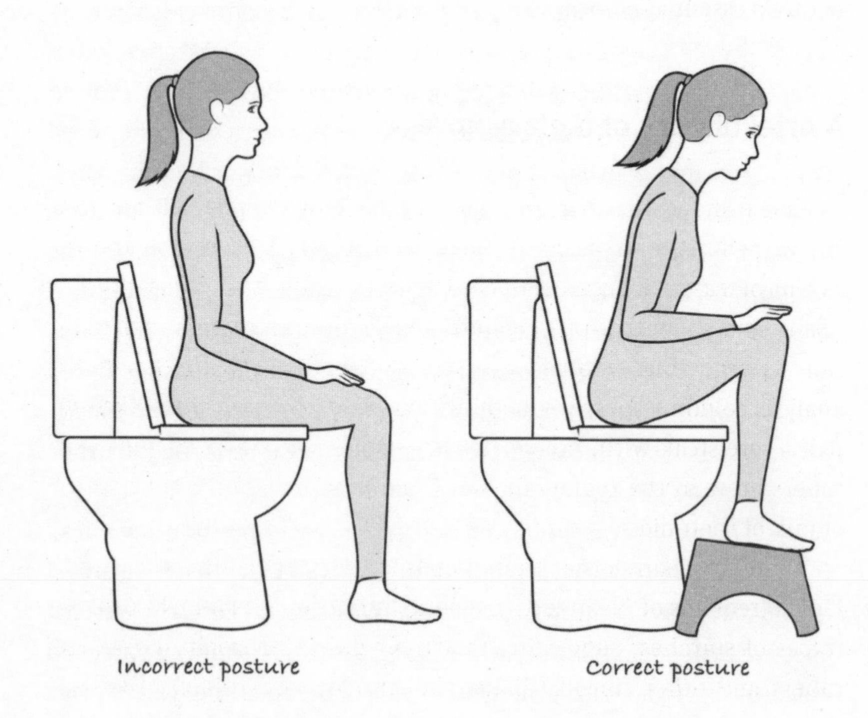

pation is through a close look at the food we eat. Dietary fibre, or to be more specific insufficient dietary fibre, is a huge contributing factor to constipation, with modern diets rich in processed foods and carbohydrates sorely lacking in this vital nutrient. Often, packaging and ingredients in modern foods can be so misleading that we don't realise how much our bowels are lacking in dietary fibre until our diets are scrutinised by a specialist.

Dietary fibre is currently in the midst of a revival as the scientific community has once again rekindled an interest in it. There was once a time when gastroenterologists thought fibre was an inert substance, its benefits amounting to nothing more than adding bulk to stool, and maybe helping to propel it through the gut. But this notion seems to have been turned completely upside down in the last decade or so. Dietary fibre is not just nature's broom, it's also nature's filler and nature's essential helper.

A brief history of dietary fibre

Evidence shows that our ancestors ate far more fibre than many of us ever will. Poo fossils dating back 50,000 years, otherwise known as coprolites, have been discovered in archaeological digs in southern Spain, and show that our Neanderthal predecessors ate a variety of fibrous vegetables and plant-based foods. Although the chemical analysis couldn't specify exactly which plants Neanderthals liked to pair a rare steak with, pollen analysis suggests that berries, nuts and tubers grew in the region and so would have made up a substantial chunk of their diets.

There are many other examples from our prehistoric ancestors. Dental remains of Neanderthals found in Iraq and Belgium also show traces of starches, suggesting that their diets were rich in grasses, tubers and other fibrous plants. In Antelope Cave in north-west Arizona twenty human coprolites were found, showing that the inhabitants ate flour made from maize, wild sunflower and other seeds, as well as fibrous succulent plants like yucca and prickly pear.

This diet was much higher in fibre than anything you or I eat today. In fact, the fossilised poos found in Antelope Cave were around three-quarters fibre by volume, which indicates that they might have been eating some 200–400g of fibre a day. (To put that in perspective, today's guidelines recommend 30g of fibre a day for adults; or, to put it another way, 300g fibre is like eating thirty bowls of Shreddies *per day*.) These findings certainly challenge the formerly popular view that our Neanderthal ancestors became extinct because they depended too much on animal protein. The suggestion now is that that they were much more omnivorous and less carnivorous than previously thought.

Fast-forward a few thousand years to the 1830s where Presbyterian minister (and advocate for dietary reform) Reverend Sylvester Graham decided that bran was the cure for a poor diet. Now, the dear Reverend *hated* commercially produced flour and breadmaking. He felt that in its whole state, wheat was a God-given source of nourishment, whereas the processed version used to make bread flour had been 'tortured' and separated from its nutrients into 'miserable trash'. (What he would make of a Big Mac is anyone's guess.)

Graham responded by creating his own form of coarse, unsifted, fibre-rich wholewheat flour which became a dietary staple for him and his flock. You may be familiar with one of his inventions, which was used to bake Graham bread and Graham crackers. However, today's Graham crackers would probably make the dear reverend very angry indeed; the original recipe did not contain sweeteners or additives, was not used to make crunchy cheesecake bases and would most definitely not have been used to sandwich toasted marshmallows and molten chocolate into a S'more.

As early as 1954, scientific research was being carried out on the topic of fibre intake. The first to shed some light on the topic was epidemiologist Dr A. R. P. Walker, who studied the Bantu tribes of South Africa. He noticed that they pooed often, and a lot. To get to the bottom (sorry) of this phenomenon, he performed an experiment asking a number of the Bantu people to swallow radiopaque pellets (these appear white on radiographic images), and then void stool into plastic bags which he labelled with dates and times. The stool bags were then X-rayed and Walker calculated how long the pellets took to complete their voyage through the bowel. The findings were telling. The slowest half of Bantu had a transit time that was quicker than the fastest third of Caucasians who were tested in the same way. He attributed this to the fact that the Bantu diet was fibre rich with millet and corn porridge. Thus, eating 'shitloads' of fibre (excuse the pun) such as in the traditional African diet was felt to prevent constipation.

By the late 1960s, the word fibre had stopped being something that free-thinking cracker magnates talked about and, courtesy of a charismatic Irishman Dr Denis Burkitt, it became increasingly common in mainstream vocabulary. While studying the eating habits among groups of people in Uganda, Burkitt observed that middle-aged people there had a much lower incidence of diseases that were common in similarly aged people living in England. Now, we aren't talking about insignificant ailments here; his list included colon cancer, diverticulitis, appendicitis, hernias, obesity, varicose veins, diabetes, dental cavities and atherosclerosis. He tried to identify the significant difference between the two lifestyles and, in 1969, he published a controversial article entitled 'Related disease–related cause?' which claimed that a lack of dietary fibre in the British diet (and its prevalence in the Ugandan diet) was the culprit.

On the basis of his observations, Burkitt advocated that people include at least 50g per day of dietary fibre in their food intake, although intake in high-income countries is only around 15g per day. (There is about 2g of fibre in a slice of wholemeal bread and 7g in 100g mixed nuts – so he was recommending *a lot* of fibre each day!) Since Burkitt's death in 1993, a number of large-scale epidemiological studies have demonstrated its numerous health benefits.

What exactly is dietary fibre and why is it good for me?

In nutrition terms, although fibre is considered an essential part of a healthy-eating plan, it is not theoretically an 'essential nutrient'. This is because, technically speaking, fibre deficiency does not exist.

Dietary fibre is actually made up of a number of different constituents, each with slightly different biological effects. However, for most of us, it's accurate enough to divide fibre into two key groups: soluble and insoluble. Soluble fibres dissolve in water and include plant pectins and gums. In contrast, insoluble fibres (usually plant cellulose) do not dissolve in water. Most plants contain both soluble and insoluble fibre components in different amounts, but oats, beans and fruits are higher sources of soluble fibre, while wheat, nuts, seeds, bran, fruit skins and legumes are richer in insoluble fibre.

Fibre (soluble and insoluble) cannot be digested in the mouth, stomach or small bowel. Because of this, it enters the colon whole, which means that some whole fibres will become part of our poo (don't look down after eating corn or tomato skins). However, anaerobic microbes in our colon can break down some fibre, fermenting it to form short-chain fatty acids which provide energy, regulate our immune systems and are involved in the gut–brain signalling that we discussed in Chapter 8 (see pp. 226–243). Fibre also exerts a probiotic effect on gut bacteria, feeding them and helping us to cultivate a healthy microbial ecology in our nether regions.

As if this wasn't enough, fibre consumption has even been linked to a high degree of confidence, with a reduction in heart disease, Type-2 diabetes, stroke and colorectal cancer. Some studies go as far as to suggest that for every 8g increase in dietary fibre per day, the risk of total deaths and incidence of coronary heart disease, Type-2 diabetes and colorectal cancer can decrease by up to 19 per cent.

So, it seems pretty clear that fibre is good for our guts, and by extension, our whole sense of wellbeing. But what effect does it have on our appetites? Well, fibre has the ability to affect the viscosity of the stuff that moves through our gastric systems. Soluble fibres form thick, gel-like gut contents, leading to the delayed breakdown of nutrients and their absorption into the bloodstream. Therefore, the result of having a fibre-rich diet is threefold: absorption of glucose and cholesterol is decreased, the stomach empties more slowly (leaving you feeling full for longer) and your overall appetite is reduced. What this means is that glycaemic control (how quickly your food results in changes in blood-sugar levels) is optimised, courtesy of the fibre-rich foods in your diet.

Intuitively, this makes sense to many of us. A fibre-rich breakfast like Shreddies literally markets itself as the cereal that can 'keep hunger locked up till lunch', and generally speaking, it does. Seedy rye bread sandwiches certainly keep me fuller for much longer than their white-bread equivalent, due almost entirely to the high fibre content of the rye.

Given that only 9 per cent of UK adults actually meet the current guidelines of 30g fibre a day, there is clearly a lot of room for improvement! However, for anyone feeling daunted by the prospect of increasing their intake, or confused about how to go about it, there are lots of easy tips and tricks here to achieve it without it seeming a chore.

The old dogma that fibre equals bran (or other cardboardy brown food) is wholly outdated. Fibre-rich foods come in a rainbow of colours, textures and taste sensations, making cooking with them a complete joy rather than a total bore. Remember, this is one of the most powerful ways you can bring yourself one step closer to finding digestive health and happiness, irrespective of whether you have ever found yourself constipated or not. And to help you on your journey to consuming more fibre, I have designed a few fibre-dense recipes that would please even the most fibre-sceptic among you (see pp. 308–317).

A key factor to remember is that a slow and steady increase of fibre in your diet is far more likely to be successful than a sudden one which could lead to troublesome bloating, abdominal cramps or increased flatulence, all of which we are trying to avoid wherever possible.

Give fibre the promotion it deserves

Fibre is a bit like the human resources department of your body. It functions in the background, quietly making things work without too much fuss. It'll never win any glamorous awards, but without it the whole organisation would quickly disintegrate – and when things go wrong, you know who to call.

It's the same with fibre. You don't really notice it when things are working because the whole point of fibre is that it operates behind the scenes. But we need to bring it out of the back office and into the spotlight, to celebrate its work and make sure that it gets the 'Employee of the Month' award a bit more often.

In a nutshell, increasing your fibre intake it is not hard to do. However, it does require a decent amount of commitment and motivation, and you'll need to maintain the increase for a while before the benefits become really apparent. But when they do, you'll wonder how you ever lived without it.

Breakfast

One of the easiest places to incorporate more fibre into your diet is at the breakfast table. Breakfast foods are almost naturally designed in a way that their fibre content can be increased. For example, you can opt for an already fibre-dense, low-sugar breakfast cereal like Weetabix, and then add in extra linseeds and a handful of fibre-rich nuts for an even greater fibre hit. Yoghurts can easily be boosted with chopped fruits and seeds; my personal favourite is live yoghurt with figs, dates and sunflower seeds, eaten first thing at the crack of dawn, while watching the news.

Bircher muesli, invented by Swiss doctor Maximilian Bircher-Brenner, is made by soaking oats and seeds in water or milk to make them more palatable, and then combining this with chunks of crunchy apple, chopped nuts and honey. It has become hugely popular recently, but actually the original version was more apple-heavy than oat-heavy, and was called *Apfeldiätspeise*, or the apple diet dish. Either way, whether you make your bircher apple or oat heavy, it remains a versatile, cost effective and fibre-dense breakfast.

Lunch

Moving on to lunch, a few simple changes can make a huge difference. Swapping processed white bread for wholemeal or seeded wholegrain breads (or even a seedy rye bread, if you really want to push the boat out) is a great way of boosting fibre consumption. Millet flour and spelt are rich in fibre and can easily become part of your culinary repertoire – I use millet flour to make a fluffy, savoury, spiced pancake in which the batter is mixed with fibre-dense turnips and served alongside a turnip raita (see p. 314 for the recipe).

Substitute your processed orange juice for two whole oranges and swap the chocolate bar for a few dates or apricots. You could even dip them in dark chocolate if you really want to live decadently. For snacks, swap crisps for popcorn, a surprisingly good source of dietary fibre: my recipe for Bombay-mix-flavoured popcorn that includes nuts and raisins is testament to how addictively good fibre-rich popcorn really is (see p. 308).

Dinner

To increase your fibre at dinnertime you could substitute wholewheat pasta for the normal version. It tastes virtually the same, and even your children will most likely not know the difference. Brown rice, too, is beautifully nutty and satisfying, and retains a slightly chewy texture perfect for a vegetable-packed rice salad with a punchy citrus dressing. And let's not forget about bulgur wheat, the gloriously scented grain that acts as a sponge for spice and punchy flavours and can be prepared in a variety of interesting ways. Getting fibre-dense beans into you is desperately easy given the wide availability of tinned beans. My Turkish-style bean salad, made sexy through some inventive seasonings, takes literally minutes to put together and is a fantastic method of creatively augmenting your fibre intake (see p. 311).

If you are a potato enthusiast (and who isn't?), eating the crispy baked skins instead of chucking them away boosts your fibre intake substantially, while reheating pre-cooked and cooled potatoes increases the number of resistant starches on their surface, giving you a lower sugar spike when you eat them. So, for a fibre hit, pick the potato salad over the chips (the potato salad having been cooked and cooled, so developing resistant starches).

I stock up my freezer with frozen vegetables, and the cupboard with tinned lentils or chickpeas. I urge you not to restrict yourself to thinking that lentils are only for making Indian-style dahl. They are celebrated by many food cultures globally and having a few different lentil recipes under your belt is well worth the effort. I have included my recipe for Middle Eastern 'salata adas'-style lentils with duck breast (see p. 312) – a properly grown-up dish that you can cook to entertain guests, while also relishing the fact that you have improved fibre intake for your diners.

When it comes to dessert, there are some tips and tricks that can help you incorporate more fibre here, too. You can opt for tinned fruit instead of ice cream one night (slippery tinned peaches never lose their appeal for me). Some fruits are particularly fibre-dense, like raspberries, apples (with the skin on), mango, passion fruit, persimmon, guavas, pomegranates, bananas, oranges, blueberries and strawberries. Incorporating generous portions of these fruits into your desserts may be the way forward. How about doubling quantities of fruit in your trifle or clafoutis, or topping frozen yoghurt with these fruits and seeds to create an instant fruit bark? I have shared my recipe for Persimmon Frangipane Tart with you on p. 315, but you can use the same recipe to load any open puff-pastry tart with even more fibre-rich fruit than I have used.

Small changes like the ones I have described above will add up over time, and can, if you stick to them, make a huge difference to your overall health and wellbeing. They will also help you poo freely and in large volumes – and who among us does not enjoy a free-flowing, large-volume poo?

Kiwis to the rescue

No, not New Zealanders (although they are very lovely people, I'm sure). I'm talking about the kiwi fruit, the berry of the Actinidia vine. It comes in two varieties, Hayward green (*Actinidia deliciosa*) and gold (*Actinidia chinensis*). Two kiwi fruits, together weighing approximately 300g, will contain 12g of fructose, 12g of glucose and a huge 9g of dietary fibre (of which a third is soluble and the remainder is insoluble). Even better, the cell walls of kiwi fruits are sponge-like, with huge water-retention capacity. They can swell to over three times their original volume – that's up to six times more than an apple and double the amount of swell caused by the laxative psyllium (an over-the-counter remedy for constipation).

These properties make kiwi fruit an extremely helpful natural remedy for the treatment of constipation. When healthy volunteers were asked to eat kiwi fruit before scanning their bowels in an MRI machine, the results showed that the fruit drastically increased water retention in the participants' small bowel and colon. Increased water retention in the bowel means, in real-world terms, more frequent visits to the toilet, with looser consistency of stool. As a result, the growing consensus in the scientific community is that these properties make the kiwi fruit the perfect natural alternative to laxatives in most mild to moderate constipation states. If you struggle to cook with them, why not try them as part of a scented citrus and kiwi platter for a light, refreshing summery dessert (see p. 317).

Prunes

Prunes are essentially dried plums, dehydrated in hot air at 85–90 degrees for just under a day. They have, sadly, been subject to bad press over the years, somehow evading the sexy culinary status that dried apricots, raisins or figs seem to have acquired. Personally, sexy status or not, I love the sweet, earthy, jammy, treacle-like taste of prunes, and often pair them with lamb or use them to make sweet treats. They even have a sort of spiritual home: Yuba City in California, USA, is considered by many to be the prune capital of the world, and historically, an annual prune festival was held there.

A hundred grams of prunes a day will give you around 6g of fibre a day. But not only are prunes high in fibre, they also contain sugars like sorbitol and phenolic compounds, which are thought to exert an array of beneficial effects on our digestive systems.

Although it's no secret that prunes possess a fairly active laxative effect, it turns out that studies on the effects of prunes on the gut remain fairly scant. What evidence there is, however, suggests that prunes work on the gut in a number of different ways: insoluble fibres like cellulose increase the bulk of your poo, while soluble fibres, like pectin, are fermented by colonic microbiota, feeding and proliferating your gut's bacterial populations, leading to the production of those beneficial short-chain fatty acids and consequently increasing, you guessed it, the bulk of your poo. Overall, the suggestion is that 100g of prunes will probably be as effective as the laxative psyllium (if not more so) and will improve both the consistency and frequency of your poos.

If you struggle to eat prunes whole or in juice form, my recipe for Prune, Linseed, Earl Grey and Cardamom Buns will be just the thing for you (see p. 309). Linseeds, also known as flaxseeds, pack a massive 27g of fibre per 100g of seed, making them the perfect fibre-rich partner for prunes in this artisan bread.

Coffee

After drinking a cup of coffee, around 30 per cent of us will feel the need to poo. For some, it can just be the smell of coffee that triggers a rumble in the bowels, and in some particularly sensitive souls, even a picture of it can get their guts going. In fact, in patients who have suffered spinal injuries, or who have somehow severed the nerves required to open their bowels, a hot cup of strong coffee can stimulate bowel action quite effectively.

For many years, it was thought that caffeine was responsible for the laxative effects of a cup of coffee, but over time, we've noticed that decaffeinated coffee had similar effects, while caffeinated soft drinks did nothing to stimulate our bowels. Coffee is a concoction of literally thousands of different chemical compounds, so working out which one causes laxative effects can be tricky. It is now thought that the acidic nature of coffee may lead to our stomachs secreting more gastric acid than usual, causing them to dump their contents more quickly into our intestines. This could help explain why coffee seems to accelerate the digestive process in some people.

But coffee also has an impact on the colon, particularly the last segment lying on the left-hand side of our abdomens. Coffee is thought to increase the levels of a number of hormones released into the bloodstream, one of which is called gastrin. Gastrin is in charge of initiating the peristaltic movements of the bowel described earlier, and also triggers what is known as the gastro-colic reflex – a physiological reflex that controls the colon, causing it to contract in response to the stomach stretching. If you have had a baby, you may have observed that they can sometimes soil nappies at the same time as they are feeding; this is the gastro-colic reflex at work.

The idea that things to do with poo are somehow taboo, unsuitable for public discussion, is probably one of the most heartbreaking things for me as a gut doctor. Overcoming this barrier and fostering a healthier relationship with your poo, which is influenced so intimately by what you eat, is a necessary step towards achieving digestive health and happiness.

In a way, talking about being at peace with your body and happy with your relationship with food, without understanding or even being able to talk freely and candidly about what happens when things go wrong (e.g. constipation) is like trying to understand world politics, but refusing to talk about war because it is an unpleasant topic of conversation.

It's true, war might not be the nicest subject to talk about, but you're fooling yourself if you think you will be able to gain a holistic understanding of world politics without discussing it. And similarly, you're kidding yourself if you think that you can gain a deeper understanding and sense of calm about your body's relationship with food without talking about what happens when you poo (or can't poo, for that matter).

Summary

- Constipation affects a large portion of the global population and is multifactorial in nature. Poo shame is very real; many people feel embarrassed addressing their constipation.

- Low-fibre diets contribute to constipation. British guidelines currently recommend 30g of fibre a day and most people fall significantly short of this target.

- Fibre is not one substance, but a group of them; these cannot be digested by the small bowel and enter the large bowel, where they are broken down in part by gut bacteria.

- Increasing fibre intake is linked to a multitude of health benefits according to evidence emerging from large-scale epidemiological studies.

- Adding 100g of prunes and two kiwis or a spoonful of linseeds to your diet is a helpful way of increasing fibre intake and addressing mild constipation states.

- Fibre-rich food is not all brown coloured. Fibre-rich foods taste delicious and can be creatively assimilated into the food we eat on a daily basis.

Bombay-mix-style Stove-top Popcorn

Serves 4

It may surprise you to know that popcorn is an excellent source of dietary fibre, packing a whopping 13g per 100g. It is 100 per cent wholegrain, after all. Air popping the popcorn is ideal, but I use a touch of oil and sugar here. Its addictive quality is such that you may find yourself having to prise your own hand out of the popcorn bucket. And sharing it with a loved one proves to be problematic, too.

Ingredients

100g corn kernels
1 tablespoon vegetable oil
1 teaspoon red chilli powder
1 teaspoon chaat masala
½ teaspoon table salt
150g roasted salted peanuts
150g raisins
100g caster sugar

Method

1. Place the corn kernels and vegetable oil in a large heavy-bottomed saucepan. Cover and heat over a medium heat. Wait for the kernels to pop; you will hear them popping away after a minute or two. Shimmy the pot around every 30 seconds or so to distribute the heat and help the kernels pop. When the kernels stop popping, or have reduced popping significantly, remove them from the heat and carefully tip them into a large serving bowl. Scatter over the chilli powder, chaat masala and salt. Roughly chop the peanuts and raisins and add them to the popcorn.

2. Heat the caster sugar in a pan until it melts to form a deep golden brown caramel. Working quickly, transfer the seasoned popcorn, raisins and nuts to the caramel and stir well with a wooden spoon. The idea is that the caramel coats as much of the popcorn as possible and makes the popcorn, raisins and nuts stick

together in little clusters. Spread the popcorn out on a wide, flat tray and leave to cool slightly before serving. The caramel will have made it extra crunchy, and the juxtaposition of sweet, salty and spicy flavours is rather spectacular.

Prune, Linseed, Earl Grey and Cardamom Buns with Prune Butter

Makes 8 buns

Both wholewheat flour and prunes are an excellent source of dietary fibre and a natural remedy for those of us who find ourselves constipated. These buns are a great treat for breakfast and will convert even the most hardened prune hater into a dried-plum enthusiast. The prune butter, sometimes caller 'lekvar' in Eastern Europe, is more akin to jam, and although named 'butter', it contains no fat whatsoever.

Ingredients

250g pitted dried prunes
1 Earl Grey teabag
250g strong white bread flour
250g wholemeal flour
100g whole golden linseeds
½ teaspoon salt
½ teaspoon sugar
Seeds of 6 large cardamom pods, crushed lightly in a mortar and pestle
1 x 7g sachet yeast
375ml hand-hot water

For the prune butter:
400g prunes, roughly chopped
700ml water
50g soft light brown sugar

Method

1. Preheat the oven to 180°C fan.

2. Start by placing the prunes in a bowl, together with the teabag, and cover with tepid water. Allow the prunes to soak for 30 minutes.

3. Add the flours to a mixing bowl along with 80g of the linseeds and the salt, sugar, crushed cardamom seeds and yeast. Mix everything together thoroughly with a spatula. Make a well in the centre and pour in the water. Mix together gradually to form a dough; the precise amount of water required will depend somewhat on the type of flour you use.

4. You're aiming for a smooth dough that leaves the side of the bowl clean. A slightly wet dough at this stage is probably better than a dry, firm one. Transfer the dough to a lightly floured surface and begin to work it. The idea is to develop the gluten in the flour as this provides it with structure. Try not to pummel it with the heel of your hands; rather, slide your fingers underneath it like a pair of forks with your thumbs facing upwards, and bring it back on itself, away from you, on to the kitchen counter. This will incorporate maximum air into the bread.

5. After 5 minutes of kneading in this way, drain the prunes and discard the teabag. Chop the prunes into quarters and work them into your dough, so they are spread through it evenly. Shape the dough into eight equal-sized dough balls and place them on a lightly floured baking try. Sprinkle over the remaining 20g of linseeds. Cover with lightly greased clingfilm and allow to rest for around an hour, or until the buns have doubled in size. Transfer the buns to the oven and bake for 25–30 minutes. You can tell they are ready when they make a hollow noise on being tapped underneath. Remove from the oven and cool on a wire rack.

6. For the prune butter: place the prunes and water in a saucepan and bring to the boil. Allow the prunes to soften completely, simmering uncovered for around 20 minutes. When the mixture has mostly dried out and only 3–4 tablespoons of water remain in the saucepan, stir in the brown sugar until it has dissolved completely – this takes around 3–4 minutes. At this point you have a choice: if you prefer a coarse texture, mash the contents of the saucepan with a potato masher or a fork. If you prefer a smooth, purée-like consistency, use a blender to achieve the desired texture. Cool the mixture before transferring to a sterilised jar and store any leftovers in the fridge.

'Toot' Bean Salad

Serves 2 (with some left over for a lunchbox the next day)

Lovingly named 'Toot' salad by my family (given the gas-forming properties of beans), the seasonings here transform the unfashionable drab tins of mixed beans into something quite fresh-tasting and spectacular.

Ingredients

480g mixed beans, drained (2 standard-sized tins, drained)
1 medium-sized cucumber, deseeded and diced
2 spring onions, finely sliced
200g cherry tomatoes
250g roasted red peppers from a jar
25g flat-leaf parsley
Juice of 1 large lemon
2 tablespoons extra virgin olive oil
1 tablespoon pomegranate molasses
1 teaspoon red chilli flakes
½ teaspoon dried oregano
½ teaspoon dried mint
Salt, to taste

Method

1. Place the beans in a wide mixing bowl, along with the cucumber and spring onions. Quarter the cherry tomatoes and roughly chop the red peppers and parsley, adding them to the mixing bowl and stirring well to combine. To dress the salad, add the lemon juice, olive oil, pomegranate molasses, chilli flakes, oregano and mint to the beans. Season liberally with salt, stir well and serve.

Note: don't be put off making this salad if you are short of a few ingredients. Equally, you can mix and match with what you have in your store cupboard, e.g. tamarind can be used instead of pomegranate molasses, and feta cheese and chopped pistachios would make great additions as well.

Duck with Salata Adas Lentils and Golden Sultanas

Serves 4

There is something quite grown up about this dish. The sort of food you might make for a dinner party with friends. Salata Adas is a Middle Eastern garlicky lentil dish. Its nutty, herbal notes are perfect against the sweet pink flesh of duck. Green lentils are, of course, a rich source of dietary fibre.

Ingredients

250g dried green lentils
4 duck breasts with skin on, excess fat trimmed
75g butter
3 garlic cloves, finely sliced
5 anchovies (from a tin)
1 heaped teaspoon allspice
1 teaspoon red chilli flakes
Juice of 1 large lemon
30g chopped flat-leaf parsley
50g golden raisins, soaked in warm water for 30 minutes, then drained
Sea salt flakes

Method

1. Start by preparing the lentils according to the packet instructions; they will need around 20–30 minutes of boiling in plentiful water. Try not to overboil and make them mushy, as this will give the final dish a soupy texture. It is better to keep them firm but cooked through at this stage. Drain the lentils and set aside.

2. Bring the duck out of the fridge and allow it to reach room temperature. Score the breasts in a criss-cross pattern through the skin, being careful not to cut all the way to the flesh. Season liberally with flaky sea salt.

3. Place the breasts skin-side down on a cold frying pan and gradually heat the pan; this allows some of the fat to slowly render and produce a lovely, crisp duck skin. Keep frying the duck in the fat you have rendered until you have melted as much of the visible white fat from the skin as possible, and the skin is a deep, golden, crispy brown colour. This can take around 10–15 minutes. Turn the breasts and cook for a further 3–4 minutes until the meat is browned all over. Remove the duck breasts from the pan and allow them to rest while you prepare the beans.

4. Melt the butter in the rendered duck fat and once it is bubbling, add the garlic and anchovies. When the garlic is golden brown and the anchovies have melted, toss in the lentils. Stir the fat through the lentils and add a splash of water followed by the allspice, chilli flakes and lemon juice.

5. To complete the dish, turn off the heat, check the seasoning and toss the parsley through the lentils. Slice the duck breasts to around 1cm thick, working with rather than against the grain of the meat. Scatter the lentils on the bottom of a large serving platter and top with the slices of pink duck. Scatter over the plump golden raisins and serve.

Millet Flour, Spring Onion and Turnip Pancakes with Turnip Raita

Serves 4

Millet is a cereal grain that belongs to the grass family, Poaceae. It is one of the earliest cultivated grains and thousands of varieties exist around the world. Its flour is fibre-dense, mild, nutty and sweet. It can be made into flatbreads on its own, but the texture can be rather short, which is why this recipe, being a cross between a flatbread and a pancake, is the perfect treatment for millet flour. Brunch awaits.

Ingredients

600g turnips
150g millet flour
100g plain flour
1 teaspoon bicarbonate of soda
1 teaspoon baking powder
2 medium eggs
1 teaspoon turmeric
1 teaspoon red chilli powder
1 teaspoon cumin seeds
4 spring onions, finely chopped
2–4 green chillies, finely chopped
500g natural yoghurt
250ml warm water
1 teaspoon chaat masala
Vegetable oil, for frying
Salt, to taste

Method

1. Peel the turnips and chop them into large, but equal-sized chunks. Boil in salted water for around 15 minutes, or until a knife just penetrates the flesh easily. Drain and leave to cool completely. Grate the cooked turnips and set aside.

2. Now make the pancake batter by combining the two types of flour, bicarbonate of soda, baking powder, eggs, turmeric, red chilli powder, cumin seeds, spring onions, green chillies, 100g of the yoghurt and a third of the grated turnips in a large mixing bowl. Beat with a wooden spoon to combine and add just enough

water to form a thick batter (I tend to need just shy of 250ml). Season liberally with salt.

3. To make the pancakes, drizzle a teaspoon of vegetable oil on to a non-stick pan over a medium-low heat and drop a large ladleful of the pancake batter into the centre of the pan. When bubbles appear on the surface it is time to flip the pancake with a palette knife. They usually take 2 minutes each side to cook through. Continue preparing the pancakes until you have used up all of the batter.
4. Make the turnip raita by mixing together the remaining yoghurt with the remaining turnips and chaat masala.
5. Serve the pancakes with the turnip raita and devour immediately.

Persimmon Frangipane Tart

Serves 4

Persimmon, otherwise known as 'Sharon fruit' is a fibre-dense, yellowy-orange fruit, resembling a tomato. Its glorious honey-sweet, slightly tannic taste even led the ancient Greeks to call it 'the fruit of the gods'.

Its popularity has increased in recent years and it can now routinely be found on our supermarket shelves. I have enjoyed it many times as a sweet addition to salads with cress and a hint of lime juice, or with feta cheese, pickled chillies, coriander and mint. But given the tart treatment, paired with almonds in the form of frangipane, is where the persimmon truly shines.

Ingredients

1 sheet puff pastry
3 persimmons, sliced into 1cm thick rounds
1 heaped tablespoon caster sugar
2–3 tablespoons apricot jam
Handful of flaked almonds, toasted

For the frangipane:
50g butter
50g caster sugar
1 beaten egg
60g ground almonds
1 heaped tablespoon plain flour

Method

1. Preheat the oven to 180°C fan. Lay the puff pastry on a flat baking tray lined with greaseproof paper. Score the pastry 2cm from the edge with a sharp knife, making sure not to cut all the way through.

2. To make the frangipane, cream the butter and sugar together with a wooden spoon till light and fluffy, then add the egg, beating thoroughly to combine – don't worry if the mixture looks slightly curdled at this stage. Add the almonds, followed by the flour and fold everything together to combine.

3. Using a palette knife, spread the frangipane all over the pastry, avoiding the scored border. Lay the persimmon rounds on the frangipane neatly, overlapping them slightly. Sprinkle a heaped tablespoon of sugar over the persimmon. Transfer to the oven and bake for 25–35 minutes, or until the frangipane is golden and the pastry edges are a deep golden brown colour.

4. Place the apricot jam in the microwave for 20 seconds. This will make it looser in texture. Gently brush the jam over the persimmons to create a glaze. Transfer back to the oven for just 2 more minutes; this will allow the glaze to stick to the fruit evenly. Remove the tart from the oven and cool before serving, cut into individual slices. You may wish to top with extra flaked, toasted almonds.

Scented Kiwi and Orange with Cinnamon, Honey and Orange Blossom Water

Serves 2

As previously mentioned, two kiwis a day may assist those who find themselves constipated – although from a culinary standpoint, the kiwi fruit is not really considered a trendy or in-vogue ingredient, being mostly limited to smoothies and yoghurts or frozen into ice lollies. The addition of citrus, orange blossom water, honey and cinnamon, however, makes the kiwi fruit sing in this perfumed fruit-salad-style dessert.

Ingredients

4 kiwi fruits
2 large oranges
1 tablespoon runny honey
½ teaspoon ground cinnamon
1 tablespoon orange blossom water
Handful of chopped pistachios
A few mint leaves, cut into thin slices (julienned)

Method

1. Peel the kiwi fruits and slice into 1cm thick rounds. Scatter on to a flat platter.
2. Cut the tops and bottoms off the oranges and stand them on your work surface. Using your knife, cut downwards, with even strokes, slicing the skin away from the flesh. Discard the peel and remove any remaining white pith. Cut the oranges into 1cm thick rounds (rather than segmenting them) and arrange the discs on the platter with the kiwi fruit. Try to keep the orange and kiwi fruit a similar thickness and as a single layer on your platter.
3. Drizzle the fruit with the honey from a height and then sprinkle over the cinnamon. Splash over the orange blossom water, ensuring that it touches as much of the fruit as possible. To complete the dish, scatter the pistachios and the mint over the platter.

CLOSING REMARKS

I feel privileged that I have been able to share my scientific and culinary journey through the digestive system with you, but in some ways also a little sad because, for now, the journey of discovery is coming to an end. The process of uncovering the interconnectedness of the digestive system and the world of gastronomy has been joyous and transformative, bringing me a sense of deep satisfaction. I hope that you will also come away with some positive messages to implement in your day-to-day life.

Being both a doctor who specialises in the field of digestive health and a professional chef put me in the unique and privileged position to be able to write *Foodology*. I have often been asked what it is like to have two separate careers that are, on the surface, so different from one another. But the reality is that for me, the common thread that ties my chef and doctor lives together is that both are entrenched in a desire to care for others; whether you do that through prescribing a medicine or by rustling up a tasty dinner is, in some ways, secondary.

I talk about finding 'digestive health and happiness' as a way of optimising physical and mental wellbeing via food throughout this book. I know that on the face of it, digestive health and happiness may sound a little, well ... fluffy. But I can assure you I haven't based any of my writing on hypothetical principles; the research involved in curating this book has been painstaking and meticulous. I have endeavoured to provide the most up-to-date, evidence-based science in a manner that is accessible and relevant to your day-to-day life as a cook. I have also been conscious of ensuring that no claims made in this book are overstated; where the picture is grey, I have said that it's grey and not pretended that it is either black or white.

In a world where the messages about what is 'best' to eat are conflicting and sometimes overwhelming, using science as a tool to help you foster a better relationship with food and eating is enlightening. Stripping away layers of confusion and conceptualising food as an object of pleasure before anything else allows you to rise above the maelstrom. Knowing that the human body in all its complexity is actually designed to enjoy foods with their array of flavours, textures and social and cultural value is the very foundation upon which a sense of digestive health and happiness is built. Because digestive health and happiness is not a diet; it is a mindset towards food and eating.

Finding digestive health and happiness will mean different things to different people at different stages in life. So, you may find that reading this book now, and then again in a few years allows you to take different messages away, depending on the particular challenges you face at a given moment in time. The building blocks to achieving digestive health and happiness and enjoying food in all its guises are summed up here.

The following messages pertain to how the digestive system/human body is designed as a vehicle to enjoy food and will help to make you the best possible cook you can be, hence laying the groundwork for a sense of digestive health and happiness:

- Cook, cook and cook some more. We have evolved as a species who cook, and it is this that distinguishes us from other creatures inhabiting the earth. I intentionally chose to include recipes across all the chapters in this book to highlight the fact that attaining digestive health and happiness starts in your kitchen, with your knife, chopping board, pots and pans and the contents of your fridge and store cupboard. No one is 'too old' to learn to cook; and if necessary, one can always upskill, so as to be able to prepare at least a basic meal.

- Recognise that we probably learned to taste as far back as when we were in our mothers' wombs, and that pleasure and disgust are all tied up intimately with the process of eating. Establishing flavour profiles is not just a function of the taste buds in our mouth, but involves all the senses, particularly the sense of smell. The experience of food textures is intertwined with our sense of hearing and contributes heavily to our enjoyment of food. Actively using all your senses when you cook will allow you to both produce and enjoy better-tasting food.

- Understanding that umami is the mysterious fifth taste that improves flavours can help explain why you preferentially enjoy certain dishes (like a cheeseburger, for example) so much. Harness the power of umami in your cooking, using naturally occurring umami-rich foods and condiments to enhance both the savoury and sweet dishes you cook.

- The world of spices is vast. In general, many spices probably do confer some health benefits, but these are hard to pinpoint, given the paucity of research at this moment in time. Breaking down perceived barriers to using spices and experimenting instinctively with them in your food, opens your life up to an incredibly broad range of culinary possibilities.

The next set of principles provides some insight into what the science of the digestive system tells us is most beneficial for us to eat. These outline how it is possible to balance one's love for mouth-watering food with a desire to be 'healthy', rather than viewing the two as separate entities:

- Hunger is a fundamental bodily experience that we live through many times a day. Understanding that how you choose to respond to hunger pangs can influence your long-term health is one step towards being at peace with hunger. Opting for delicious foods that give a steady release of energy rather than a sudden sugar rush is beneficial.

- Realising that no one 'diet' is helpful for everyone is critical. No magic-bullet-style 'superfoods' exist for those on a mission to shed the pounds. We are inherently all created differently and probably have different bodily responses to eating the same food groups. One path to digestive health and happiness in the future may well be the development of individualised nutrition plans that account for our differences.

- Be at one with the trillions of bugs that you host; they are a fundamental part of you and can influence health in ways that you would never previously have imagined. Your gut bugs need to be cared for and cultivated – they need to be fed and nourished, just like you need to be fed and nourished. Knowing about the foods that help diversify gut microbes will help you to connect with the world that exists in the depths of the bowel. Inventive, delicious recipes that help incorporate prebiotic and probiotic foods into your daily menus may contribute to achieving a sense of gastronomic fulfilment.

- Know that the digestive system doesn't exist as a standalone entity. The gut and brain speak to one another through a variety of mechanisms and our gut microbes mediate this communication. What you eat can impact mental wellbeing and behaviour in ways that are astounding. So, find digestive health and happiness in the knowledge that what you choose to put inside you really does matter.

- Learn to accept that flatulence, bloating and constipation are experienced by each and every one of us at one moment or another. Being able to talk openly about these symptoms without a sense of shame and recognising that food has a role to play in alleviating them are critical if you want to experience digestive health and fulfilment.

I realise that the journey to digestive health and happiness is in some ways never-ending. Ever-expanding scientific knowledge is challenging some of the food- and nutrition-related misconceptions we thought to be steadfastly true in the past. Equally, in the future, as science further pushes our understanding of the impact of food on human health, we may look back and realise the error of our ways today. One could argue that digestive health and happiness actually lies on the path of discovery: learning, evolving and engaging in an active (rather than passive) process, which bridges the gap between the world of delicious food and that of scientific discovery is where peace can be found.

I am thirty-three years old and am aware that a lifetime of learning awaits me. In some ways, you could argue that I have written *Foodology* in the very infancy of my career – both as a gastroenterologist and a chef. It would be fascinating to rewrite it after another twenty-five years and see how both medical science and the culinary arts have evolved over time and where my path to digestive health and happiness has led me. I hope that medical science and gastronomy co-evolve and move in tandem with one another more than they have done till now.

Open dialogue between the different players – from doctors and dieticians to the food industry, cookery authors, gastrophysicists, chefs, etc. – is vital if we are ever to reach true gastronomic nirvana.

My final message before signing off: the love for food, cooking and eating is central to twenty-first-century life, and I feel it is the ultimate expression of self-care. And so, I wish you all the best of luck on this, your personal journey to digestive health and happiness, and I hope that you'll find fulfilment and ultimate wellbeing at every stop along the way.

Acknowledgements

A friend once told me that it takes a village to raise a child. Well, in a way writing a book is no different. And while I admit the words that you find printed on the pages of *Foodology* are mine, the manuscript would not have come together were it not for the cumulative efforts of a number of rather marvellous people.

My husband, Dr Usman Ahmed was the first person to give me the confidence to write in this genre. Without him, *Foodology* would have been no more than a feather blowing in the wind. He is my motivator, my inspiration and my life's love: my apologies if this embarrasses him, but it had to be said!

There are literary agents and then there are friends. For me, Heather Holden Brown and Elly James of HHB Agency are the latter. I am eternally grateful to them for believing so strongly in my initial proposal. Nicky Ross, the editor of this book, has championed the philosophy of *Foodology* through to its publication and it means so much that she believes not just in the ideas presented in this book, but also in my abilities as an author. Thank you for fighting my corner.

Olivia Nightingall, Becca Mundy, Caitriona Horne, Anne Newman, Sarah Christie, Elisa Wong (cover design) and the entire team at Yellow Kite books who have worked tirelessly on different aspects of this book's production. Bravo! you have no idea how much I admire all of you. I am so excited about where the next stage of our *Foodology* journey takes us.

A huge thank you to my friend Sheriar Arjani. Who knew your English degree would be so handy! I still can't believe that you took the time to edit the chapters of *Foodology* despite baby Ivy being born early. You and I both know how much value you have added to the text of this book, helping me make it infinitely more accessible for non-scientists. What I am most grateful for is the fact that through *Foodology* we have had a chance to become friends in our own right. The value of this friendship will extend far beyond the success of this book, I am certain.

Another thank you to all those around me whose advice I have been able to benefit from. Professor Stephen Roper, the NHS Gastroenterology Consultants based in North West London Deanery and my colleague/friend Dr Arun Sivananthan (for the jokes).

Finally, on a personal note, I gave birth to my second baby son in the midst of the Covid pandemic. I have written this book whilst feeding him, rocking him and cradling him. Late nights, crying baby, milk, vomit, home-schooling and *Foodology* open on my laptop; that was my life in 2020. Both my sons have carried me through the pandemic and will always remain my driving force and my motivation to make my life and work a success. I love you both.

So many have lost so much in this unprecedentedtime. The frailty of human existence has been brought to the fore like never before. I hope that this book serves as a ray of sunshine and brings some positivity into your lives. Onwards and upwards.

Index

METRIC/IMPERIAL CONVERSION CHART

All equivalents are rounded, for practical convenience

Volume (dry ingredients - an approximate guide)	Cups	Grams
butter	1 cup (2 sticks)	225g
rolled oats	1 cup	100g
fine powders (eg. flour)	1 cup	125g
breadcrumbs (fresh)	1 cup	50g
breadcrumbs (dried)	1 cup	125g
nuts (eg. almonds)	1 cup	125g
seeds (eg. chia)	1 cup	160g
dried fruit (eg. raisins)	1 cup	150g
dried legumes (large, eg. chickpeas)	1 cup	170g
grains, granular good and small dried legumes (eg. rice, quinoa, sugar, lentils)	1 cup	200g
grated cheese	1 cup	100g

Weight

g	oz
25g	1oz
50g	2oz
100g	3½oz
150g	5oz
200g	7oz
250g	9oz
300g	10oz
400g	14oz
500g	1lb 2oz
1kg	2¼ lb

Volume (liquids)

ml	oz	tsp/cup
5ml		1tsp
15ml		1 tbsp
30ml	1 fl oz	⅛ cup
60ml	2 fl oz	¼ cup
75ml		⅓ cup
120ml	4 fl oz	½ cup
150ml	5 fl oz	⅔ cup
175ml		¾ cup
250ml	8 fl oz	1 cup
1 litre	1 quart	4 cups
1kg	2¼ lb	

Oven temperatures

Celsius	Farenheit
140	275
150	300
160	325
180	350
190	375
200	400
220	425
230	450
500g	1lb 2oz
1kg	2¼ lb

Length

cm	inches
1cm	½ inch
2.5cm	1 inch
20cm	8 inches
25cm	10 inches
30cm	12 inches

Also by Dr Saliha Mahmood Ahmed

Hardback 9781473678569

yellow
kite